Education in the 80's:

SCIENCE

The Advisory Panel

Jeanne E. Bishop, Planetarium Director and science teacher, Westlake (Ohio) Public Schools

Linda DeTure, Adjunct Professor, School of Education and Human Development, Rollins College, Winter Park, Florida

Mary A. Eisenmann, chemistry teacher, Pearce High School, Richardson, Texas

Mary Jane Head, science teacher, Spring Branch Independent School District, Houston, Texas

Alice J. Moses, elementary science teacher, University of Chicago Laboratory Schools, Chicago, Illinois

Donald C. Orlich, Professor of Education and Science Instruction, Washington State University, Pullman

John J. Padalino, Director, Pocono Environmental Education Center, Dingmans Ferry, Pennsylvania

Beverly C. Stonestreet, physics teacher, Linganore High School, Frederick, Maryland

David R. Stronck, Associate Professor of Education, University of Victoria, British Columbia, Canada

Frank X. Sutman, Professor of Science Education, Temple University, Philadelphia, Pennsylvania

I. Joyce Swartney, Professor of Science, Buffalo State University College, Buffalo, New York

Education in the 80's

80's:

SCIENCE

Mary Budd Rowe
Editor
University of Florida,
Gainesville

Classroom Teacher Consultant
Wilbert S. Higuchi
Sidney High School
Sidney, Nebraska

National Education Association
Washington, D.C.

Stock No. 3161-X-00 (paper)
            3162-8-00 (cloth)

Note

The opinions expressed in this publication should not be construed as representing the policy or position of the National Education Association. Materials published as part of the NEA Education in the 80's series are intended to be discussion documents for teachers who are concerned with specialized interests of the profession.

Library of Congress Cataloging in Publication Data
Main entry under title:

Education in the 80's—science.

   (Education in the 80's)
   1. Science—Study and teaching. I. Rowe, Mary Budd.
II. Series.
Q181.E46   507'.1   81-18865
ISBN 0-8106-3162-8   AACR2
ISBN 0-8106-3161-X (pbk.)

Contents

Editor

Mary Budd Rowe is a Professor at the University of Florida, Gainesville. She has also taught at Columbia University, and served as Program Director and Section Head for the Division of Science Education Development and Research of the National Science Foundation. Dr. Rowe has acted as North American representative to the UNESCO Council on Education in the Sciences and as a member of the U.S. government delegation to China on Education in the Sciences.

Classroom Teacher Consultant

Wilbert S. Higuchi is K–12 Science Curriculum Chairperson and biology teacher at Sidney High School, Nebraska.

Science as a discipline and as a moving force in society has been praised, condemned, and/or ignored. Often, it has been misrepresented due to the nebulous line of demarcation in the minds of many between science and technology. If science teachers are to develop the talents of their students in understanding scientific concepts, utilizing scientific knowledge, and/or pursuing science careers, they must be able to adapt or adopt techniques that will produce students who are lucid, comfortable, and knowledgeable in the field.

The pendulum in science education has been swinging back and forth over the past decades in concert with societal standards. Further, as science has been praised, furiously defended, condemned, or placed on the back burner to nurture, teachers of science have found themselves in similar positions.

A compendium of thoroughly researched materials, EDUCATION IN THE 80'S: SCIENCE is an excellent resource book that has value for the science teacher whether he or she teaches in the K–12 system or in higher education. Though it does not answer all problems, it does raise questions with implied answers as to what a science teacher can do to enhance his/her own program.

Christianson's chapter places the entire science curriculum in schools in a clear perspective. First he traces examples of societal problems as they relate to daily living. He then projects the implications for the future and discusses our obligations as educators of the generations to come. Christianson's approach to science education is not of limited scope, but rather it encompasses other areas of the school's curriculum. It is an appropriate introduction to the succeeding chapters on science education.

Johnson and Johnson discuss the potential effects of student–student interaction in the science classroom. They build their case for student interaction at various levels by illustrating that teachers and students' working together heavily influences the students' learning outcomes. Competition, achievement, and attitudes are reviewed based on the research data available. It is not a simplistic approach—but thought provoking and quite comprehensive.

Teacher and student behaviors is the theme of Okey and Butts' chapter. The authors discuss many techniques that have led to improved student achievement at various academic levels. The reader is able to identify the links between teacher behavior and student achievement, and then use appropriate strategies to help his or her students improve themselves in science education.

Koran and Shafer explore the informal settings in which students receive science education. The diversity among learners in informal settings is compared to that of learners in formal classroom settings. Also, they discuss studies citing the advantages of teachers' providing more science knowledge before, during, and after student field trips or museum visits. Then, too, the authors note the advantages of incorporating certain learning methods into informal settings—duly recognizing that some comprehensive structuring is needed to fully utilize informal settings, but cautioning about pitfalls while using this method.

Bredderman presents the pros and cons of activity-based science programs in the elementary school. He does not argue for one program over another; nor does he dwell on the economics of instituting such a program in a school system. He does point out the pitfalls in the research completed thus far and leads the reader to be skeptical about the claims of certain programs. He also suggests ways in which teachers, curriculum specialists, administrators, and/or school boards can better evaluate programs while attempting to select one for adoption. Bredderman uses tables in his study to point out specific achievements and shortcomings, and he also points out the flaws in administering tests that are not in concert with the aims of the program utilized—e.g., using a content-oriented test to evaluate achievement in an activity-based program and vice versa.

Berger highlights studies based on the attitudes of students and the learning of science. The results of studies contrasting the interest levels of elementary, middle, and high school students and young adults confirm the need for astute science teachers. He also discusses teacher attitudes toward teaching science on the elementary level and raises some pertinent questions in the area of elementary teacher training programs. Do elementary teachers feel insecure with the teaching of science due to the lack of an adequate background? His emphasis of Rowe's study in "response time and success" in students' achievement in science may be a novel trend to counter some of the concerns expressed early in the chapter.

Hegarty writes a thought-provoking chapter as she discusses the role of laboratory work in college science courses. She begins by citing studies that reveal the value or uselessness of lab-oriented programs and of traditional classes with cookbook laboratory activities. Further, she contrasts the achievements of students in programs for science majors with those in programs for nonmajors. Hegarty discusses the support that research shows for a relationship between laboratory activities and learning for science majors, and alludes to the residual gains made by nonmajors. Basically, much of the research cited deals with college-level programs, but the implications and influences these findings have for secondary science programs are quite clear. Hegarty's findings on computer use, the inquiry method, and faults in the confirmatory exercises in programs may be enlightening for many and disappointing to others. Of note is her discussion of the literature on students' cognitive styles and personality types in relation to learning.

McDermott's chapter may be considered an extension of Hegarty's writing. She discusses problems in college students' understanding of kinematics and the implications of such for high school physics students. She explores the problems of misconceptions that affect students' understanding of concepts in physics classes. The comparative study of a group of students learning a specific concept provides some interesting insight into the learning–achievement ratio. The success rate for academically disadvantaged students as compared to that of other groups of students is very significant for argumentative purposes. McDermott's comprehensive review of one area of physics raises some interesting problems that educators must address in order to help students in their classes.

Minstrell raises several problems similar to those discussed by McDermott but now in relation to the secondary science program. While the results of his research with his own high school students are based on entire class production, as compared with the individual response results presented by McDermott, they may be eye-opening for many. He expands on his results, measuring sense experiences and rational argument in dealing with alternative ways of organizing phenomena into a meaningful conceptual entity. He also leaves you with ideas as to how teachers can adapt some of his techniques in order to improve teaching methods in their own classrooms. Minstrell's work does raise some questions whose answers form the core of teaching and learning . . . how best to teach so students can learn.

Lipson and Lipson discuss the role of computers in education. Their approach is reassuring even to the most "bashful" teacher who feels uncomfortable with gadgets. The authors suggest techniques or methods for using computers in instruction and show how simulation activities add to the learning process of students. They do not preclude actual laboratory experiments because one uses computers in "solving" problems, but rather they show how both can be tied together so that students have a wider range of learning experiences. In closing, Lipson and Lipson raise some interesting questions that affect teaching through the use of computers. Does their work counter the literature cited in previous chapters as to the use of computers in the classroom?

EDUCATION IN THE 80'S: SCIENCE is a book that projects the areas of concern in this discipline. The several authors offer concrete evidence as to the pitfalls and gains made in the past, and then offer recommendations for improvements in science education. The large list of references allows individuals looking for specific information to go further. The imaginative thoughts that are proffered encourage enthusiastic science teachers to adapt/adopt some of the hints to see if their students will better understand the subject matter. This book is an excellent compendium for science teachers in that many of the ideas and suggestions presented are related more to actual experiences than to theory. Science education is as important as the 3 R's: as we progress, knowledge in the field of science will help the individual to make the choices necessary for his/her survival.

Wilbert S. Higuchi
Sidney Public Schools
Sidney, Nebraska

Our disposition to plan and take action depends on what we know and how we came to know it. We apply our knowledge only to the extent that we care about the consequences enough to make the effort worthwhile. Any science program ought to encourage students to answer the following questions:

What do I know?

Do I believe it?

What must I do with what I know?

Do I know how to take action?

What are the possible consequences?

Will anybody care?

Each of the chapters in this volume has something to say about one or more aspects of the science discussion cycle illustrated in Figure 1. At each level of sophistication, from novice to expert, one must learn to move around such a cycle.

Students, as well as teachers, differ in terms of which parts of the cycle attract them most. Some people, for example, find that the practical application of new knowledge fascinates them. Others who prefer to work more abstractly focus more on understanding than on control. Some who are interested in issues and values focus on the interaction between technology and society. Whatever the primary interest of each person, it seems clear that the nature of the knowledge that is relevant and the way in which it is to be used are not the same in each phase. For this reason there are important curricular and research issues to be faced by all people responsible for the planning and the conduct of science education. The percentage of general participation in formal instruction in science and technology is falling just at a time when the labor market need for scientifically trained workers is high. Moreover, the nature of the technologically based controversies we will continue to experience requires that we increase our students' knowledge and sophistication in dealing with all parts of the science discussion cycle shown in Figure 1.

FIGURE 1
FEATURES OF A DISCUSSION CYCLE
(*New Directions for Community Colleges* 31: 31; 1980)

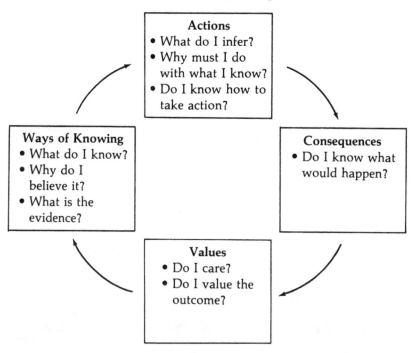

In the opening chapter, Christianson says, "What we are looking for, then, is a situation in which lay people demonstrate knowledge of both the potentialities and the limitations of science, what people in less scientific times called wisdom." How should curriculum and instruction be shaped to achieve a level of scientific wisdom suitable for participants in a modern democratic society? Christianson establishes the following curricular and research focus: to participate effectively in the public forum, students need a working knowledge of science with a procedure for updating it; an appreciation of science as a cultural discipline; and an ability to identify value commitments and to distinguish philosophical, ethical, and religious argument from strictly scientific reasoning (see Figure 1).

Johnson and Johnson, in turn, show how a science classroom can be structured to create mental and emotional conditions that would allow for pursuit of the goals described by Christianson. Both chapters focus on the part that vigorous discussion and argument will play in the development of scientific wisdom. Johnson and Johnson turn their research findings on the relationship between feelings and achievement in science into practical suggestions for action which educators at any academic level should find useful. Their recommendations have emerged from research and careful evaluation of programs in action.

In Chapter 3 Okey and Butts add to our knowledge of how to move students around the discussion cycle in Figure 1. They cite research studies that supply the basis for practical and specific suggestions regarding subject matter knowledge of the teacher, structure of the lesson, motivation of the students, rapport between the students and teachers, response conditions for the students, and physical organization of the classroom.

A great deal of science learning takes place spontaneously in settings other than the classroom. In "Learning Science in Informal Settings Outside the Classroom," Koran and Shafer report that over 300 million people visit museums each year. Half of the annual 35 million visits to science and technology centers are made by people under 18 years of age. Science apparently interests a broad spectrum of people. Through visiting museums and national and state parks, and watching science programs on television, as well as through pursuing certain hobbies, people spontaneously acquire a great deal of knowledge. What characteristics of these informal learning situations might be profitably transported into the classroom? To answer this question Koran and Shafer sharply contrast the learning and motivational conditions in formal and informal settings. They then make some interesting recommendations based on their review of research on learning under formal and informal conditions.

In these days of limited resources and a lessened priority for science, is it necessary to have a hands-on science program in the elementary school? In Chapter 5 Bredderman examines research on elementary science programs with high involvement activities and tells us who is helped most by such programs—namely, disadvantaged children. In Chapter 8 McDermott reports a similar result for college students from economically deprived backgrounds. When they participated in a largely activity-based physics course with certain characteristics, which she describes, these students performed better on final tests than did many of the physics majors. Both authors suggest that possibly many more students who are avoiding science or who have histories of poor achievement in science might be helped if their teachers were to implement some of the recommendations that they have derived from the research discussed.

What incentives are there for a student to study science? Berger tells us in Chapter 6 that the attitudes and values that people develop are closely bound to how they have acquired their knowledge. (See Figure 1.) There is a connection between an educator's philosophy and what she or he emphasizes. If, for example, the intent is to impart the maximum amount of knowledge in a given period of time, we may expect to find a great deal of rote learning being done by the student. Its incorporation into the network of relationships depicted in Figure 1 is unlikely under that condition.

Hegarty's summarization in Chapter 7 of the research on learning science reinforces the arguments of Berger, Christianson, McDermott, and Bredderman—namely, unless knowledge and process are more successfully joined than is usual in many instructional contexts, motivation as well as retention and the ability to keep on learning or changing may be seriously impaired. This would be as true for most college students as for elementary and secondary students. Hegarty's chapter should be read by teachers of both high school and college students because it provides an excellent synthesis of research on learning science.

Students at all levels seem to have more difficulty grasping physical science concepts than biological science concepts. Scores on science indicators of the National Assessment show that these difficulties begin early. Minstrell, a high school science teacher who does research on learning while he teaches, describes what may be some of the reasons why physical science concepts are so difficult to grasp. He provides examples to show that students come to class with compelling sets of naturally evolved physics concepts—many of which do not conform to current thinking in the discipline. This mismatch between what people deeply know and use as a basis for interpreting physical phenomena and the physical science concepts that educators present creates obstacles for students. Because physics ideas are often counterintuitive, Minstrell warns that simply telling students what they are supposed to think does not work. Intuitions are not so easily altered. The instructional problem is to identify the belief system that is at work in a given physics situation and then to try to engineer a strategy that confronts students with data that the existing concepts cannot encompass. He gives examples of this kind of investigation and through his own research on the job encourages classroom teachers to become investigators.

Probably Lipson and Lipson had the most difficult assignment. Since the computer era is in such transition and since readers will vary so in their background, the authors of Chapter 10 had to make some choices as to the level and the extent of their coverage of computers as instructional tools. They elected to write for teachers who are only just getting acquainted with the instructional implications of computers. As many families are investing in personal computers, more and more teachers want to learn how to use computers, if only not to be "outflanked" by their students. Lipson and Lipson argue that this technology will eventually be just as prevalent as books and the chalkboard are today. The most powerful use of computers for instruction is for simulation of complex processes or systems. The attraction of people to computers when they are placed in libraries and in museums and their ability to keep people interacting with them mean that their potential for informal education is also enormous. The instructional potential of this technology has not yet been fully realized at any academic level. It is a good time for people who are looking for a change and who are mindful of the growing "electronic culture" among the young to take a hand in shaping its future.

I have been asked to indicate which chapters are more suitable for teachers at different academic levels. This is difficult because much of the research described has application at more than one level. Certain chapters may be particularly valuable for specific interests, as follows:

	1	2	3	4	5	6	7	8	9	10
Elementary	X	X	X	X	X					X
Secondary	X	X	X	X		X	X	X	X	X
College	X			X			X	X	X	

Research can certainly provide some guidance for those who want to improve the extent and quality of student participation in science and technology. The contributors to this publication invite the reader to make applications of the research presented herein. There are implications both for teaching and for the development of instructional materials in each of the chapters.

Mary Budd Rowe

CHAPTER 1

Understanding Science As a Cultural Phenomenon—Mission for the 80's

Drew Christianson

SCIENTIFIC HUMANITIES

What we want most of all in terms of science literacy is an educated public that shows a sense of proportion about the scientific endeavor. If we are scientists or science educators, we are worried about the ups and downs of public opinion toward science, and we are concerned about the threat that these fluctuations of favor pose to the integrity of science and to the support for science instruction. What we are looking for, then, is a situation in which lay people demonstrate knowledge of both the potentialities and the limitations of science—what people in less scientific times called wisdom. After all, the hysterical fear and the exaggerated hope shown by the public result not so much from too little specialized knowledge on their part but more from too little understanding of the human side of science.

The human side of science is not a catch phrase for the failures and mistakes of scientists or for the lethal potentialities of technical progress (a veiled reference to thalidomide, Hiroshima, and Love Canal), although it encompasses these, too. Rather, more broadly, the human element refers to all the unexamined forces energizing and shaping science from within and from without. It is private ambitions, corporate motives, and public goals; it is lofty purposes and public expectations; it is group loyalty and professional allegiances. Some of these tendencies are quite conscious and publicly shared, such as the desire to eradicate polio and smallpox. Still others are personal, but self-conscious, such as the ambition of Watson and Crick to win the Nobel Prize. Finally, some, like Edward Teller's preoccupation with constructing a thermonuclear weapon, seem to escape the explanations given for them. Furthermore, with most scientists working either for industry or for the defense establishment, scientific developments are shaped by powerful values other than those of science itself. Long before science becomes a concern of the general public, it is already alloyed with other values and interests: the profit motive, economic productivity, technological superiority, and national security, for example. Accordingly, to appreciate the place of science in civilization, men and women must come to understand science as a human activity, and, more than that, they must see the practice of science as a metaphor for the human condition. They need to understand that science is an activity of fallible human beings who are inspired by noble ideals, cursed by blind ambition, and often fated to endure consequences they never foresaw.

EDUCATIONAL IMPLICATIONS

As a classroom goal, wisdom about the scientific enterprise translates into the ability of the student to interpret science and its achievements and failures in terms of the human and social forces that have formed them. Scientific humanities, to adapt a phrase from medical education, hold vast potential in this direction. History and biography, for example, offer an excellent way for the lay person to enter scientific culture. What more pleasant way to learn about Darwin's discoveries than by reading his own account in *Voyage of the Beagle* or a literate history like Loren Eiseley's *Darwin's Century*? What better way to learn about the place of colleagueship and poetic imagination in the unfolding of atomic physics than by reading Werner Heisenberg's *Conversations*? Is there a more exciting story of vanity and ambition in the service of science than James Watson's *Double Helix*? Learning science from sources such as these is both entertaining and informative. More importantly, it gives the reader firsthand testimony of the personal and social dynamics of scientific research, providing the kind of background that will help him or her evaluate the claims and counterclaims of rival parties in contemporary controversies over science and its uses.

The examination of scientific controversies, especially those that have taken place since World War II, can open up for the student a new world of understanding about the scientific enterprise as it operates today; the roles of government, industry, and the military in shaping science policy; the impact of institutional allegiances and friendships on expert judgment; the usefulness of publicity in shaking loose set opinions; and the enduring importance of the committed scientist. Some may be squeamish about exposing the vulnerabilities and vanities of the scientific community to students. Others may object with vehemence that scientific disputes and open controversy are not typical of science. To these adversaries, we should respond that ideal science is not science as it exists today, with scientists organized in large teams with huge grants from foundations, government, and the military to research problems that serve corporate and governmental purposes. The science that the student will meet as a citizen is the science of the headlines, and what he or she learns about science in newspapers will probably be controversial. Studying the scientific disputes of the recent past may assist the student in interpreting newspaper and media accounts judiciously.

The examination of scientific controversies—e.g., nuclear energy production, the use of pesticides, and the regulation of recombinant DNA research—offers opportunities for interdisciplinary collaboration among science teachers and educators in other fields such as social studies and language. Take, for example, the debate some years ago over the development of a supersonic transport (SST), an example we shall take up at greater length later. In an interdisciplinary setting, science teachers might discuss the engineering and scientific aspects of supersonic flight; social studies teachers might examine the politics involved in winning support for a major technological project, including the role of scientific experts in the advocacy of public programs; and English teachers have an opportunity to illustrate good and bad argument and to give a lesson in journalism, exploring superior and inferior examples of science reporting. Science literacy, therefore, offers a double challenge to the science faculty. The first is to give the average student a working knowledge of science. The second is to engage other educators in an effort to communicate to students the intellectual tools they will need to appreciate science as they will know it—*not as a strict discipline, but as a cultural phenomenon*, made up of news reports, political disputes, technical papers, and imperfect solutions.

VALUES AND ETHICS

A valuable intellectual skill in the interpretation of science is the ability to identify value commitments in various disciplines and to distinguish philosophical, ethical, and religious argument from strictly scientific reasoning. Some fields seem to lend themselves naturally to divisions along philosophical and ethical lines. Genetics, for example, is one area of science that has long been the subject of factional disagreements along value lines, with humanitarian defenders of individual rights opposing eugenicists, and a traditional medical ethic of patient care at odds with the ideals of preventative public health care. Demography is another obvious example of a discipline subject to ideological tensions, with technicists seeing fertility decline as a function of contraceptive engineering and those with socialist leanings attributing the drop-off to changing socioeconomic conditions. To be literate scientifically, therefore, will mean, among other things, having an acquaintance with value conflicts in the sciences and acquiring a sixth sense for the values and loyalties of scientific experts.

Nonetheless, we should not assume that our skills at spotting value conflicts will provide panaceas for all our painful problems. Controversial issues are usually made up as much of philosophical commitments as they are of scientific and technical data. Labeling one scientist as a member of one group and another as belonging to a second group will do nothing to resolve the serious disputes confronting the public. Neither will it suffice to have people identify their own preferred values, as is done in values clarification, because private choices do not provide the basis for a valid public consensus. Rather, the scientifically literate lay person should be able to scrutinize his or her own values in terms of the competing principles bearing on public issues, and so be capable of entering into a critical dialogue over them. It has become increasingly important to have open discussion before the government, universities, corporations, or research groups initiate significant new programs or even continue some old ones. It is equally important to conduct such dialogue in a reasoned manner. To this end, science literacy must also include some acquaintance with modes of moral reasoning and with developments in bioethics and the ethics of the other sciences.

MORAL REASONING

Disputes about science and its applications, like differences over other matters of public importance, are carried on in the give-and-take of public argument. Each side advances what it takes as "good reasons" to adopt a particular course of action, defends its cause against opponents' charges, and exposes the weaknesses in the adversaries' position. A good argument about the public uses of science stimulates a serious listener to examine its claims thoughtfully and makes a critic feel obliged either to respond with a careful counterargument or to concede a point. The SST controversy provides a straightforward example. American aerospace interests wanted the United States to develop a supersonic transport to compete with the Anglo-French Concorde. Critics of government support for development of the plane charged that distressingly loud noise would be produced over wide segments of the country under the plane's flight path. This was an argument that carried a great deal of weight with a public already disturbed by the roar of subsonic jets. In reply, the advocates of the SST responded that technology could be found to reduce the sound to tolerable levels. It turned out, however, that the weight of the noise suppressors would have equaled the transport's projected payload, an irony that eventually doomed the project. The argument was lost when no adequate answer could be given to a serious objection brought by the SST's opponents.

The exchange and testing of reasons is an intellectual skill that students need to acquire if they are to be capable of evaluating controversies over science and technology. Reasoning in this way is a lost art. Very often even bright students, influenced by the dramatic reporting of television and newspaper headlines and awed by the sheer volume of information they are confronted with in their investigations, know only how to report on contending views as a clash of opinions; they are unable to draw a conclusion as to which side of the argument makes the most sense. If they do draw a conclusion, it is not one that they can follow back to its premises. Rather, it is an intuitive judgment based on sentiment or ideological commitment. The purpose of teaching moral reasoning (or public argument), therefore, is threefold: (1) to link facts, values, and practical conclusions—that is, to make judgments about policy directions; (2) to test the conclusions we ourselves come to, to see whether they are justified by the facts and by the values or moral principles that we claim govern our judgments; and (3) to convince others of our judgments and so work toward a public consensus.

Implicit in the SST debate were evaluative (normative) standards, which were used both to justify and to assess claims made on both sides. In moral argument, one justifies one's own judgments by appeals to moral principles or values that either are commonly held or would readily be accepted by any reasonable person who considered the matter; then one invites others to weigh these justifications. In its simplest form, the argument made by opponents of the SST can be broken down into a traditional practical syllogism:

Major Premise (moral principle):   It is wrong to harm others.

Minor Premise (factual statement):   The SST will harm people and property.

Conclusion (practical judgment):   Therefore, it is wrong to develop the SST.

The major premise states the principle governing the evaluation of the facts. Often, of course, the moral principle will only be tacitly invoked. It is not obvious, for example, that noise is a morally significant matter. Isn't it simply an annoyance, a disamenity? Only after reviewing the impact of the sonic boom does it become clear that noise is more than an inconvenience. It then seems appropriate to categorize noise as a kind of harm and invoke the most basic of ethical principles: "Before all else, do no harm!" Sometimes the moral principles involved can be discovered only after reviewing the arguments on both sides. As a result, one task for teachers is to show students how to distinguish moral appeals inherent in certain arrangements of facts.

The minor premise states the facts as an instance (or not) of the pertinent principle—in this case, harm—depending on the findings of empirical investigation. Will windows be broken? Will real estate values fall? Will people go deaf and suffer psychic distress? And so on.

Finally, since the facts are confirmed and the noise will do harm, it follows that it is wrong to develop a plane that would do such extensive damage. If the facts had not been confirmed, then the opposite conclusion might be drawn—i.e., it would follow that it is not wrong to develop an SST.

Supporters of the SST project could respond to their opponents' argument in one of two ways. Either they might question the opponents' premise, or they might attempt to refute their statement of the facts. In this case, advocates of the SST might challenge the belief that it is always wrong to cause harm, or they might question whether the SST will, in fact, do the damage its critics claim it will. We shall look only at the first strategy, the normative one, with the hope that it will illustrate the nature of extended moral argument.

To upset the moral premise of their critics, the SST proponents might argue, for example, that it is not always wrong to do harm. Somebody is always bound to suffer to pay the price for progress, they might contend. In a formal way, they might propose another norm that would take priority over the harm principle—namely, when the welfare achieved through a project is greater than the harm inflicted, then it is permissible to carry it out. We might call this the social welfare principle. If both parties agree to the social welfare principle, then the real test would be an empirical one, assessing cost and benefits. This argument would parse itself out this way:

MAJOR PREMISE

Opponent	but	Advocate
It is always wrong to do harm to others.		It is right to increase human welfare when the good done outweighs the harm suffered.

MINOR PREMISE

To be determined: Does the SST produce an increase in human welfare?

The opposing groups would then have to consider whether the good done (decreased flight time, aerospace jobs, national prestige, etc.) would outweigh the harm done (fractured windows and eardrums, decline of neighborhoods, disruption of life, etc.).

But the critics of the SST might not want to let things get this far. They might instead want to question the social welfare principle. In such an event, the critics would then propose still another more refined principle. They might argue, for example, that it is unjust for a minority to gain an advantage at the cost of injury to a majority. In the name of fairness (justice), they might argue further that it is wrong to impose substantial suffering on anyone for the sake of trivial gains. They would then be making a *natural rights* argument that some goods are too precious to trade away. Again, the argument would proceed first at the level of principle and then move on to the determination of the facts. It would look like this:

MAJOR PREMISE

Advocate	**but**	*Opponent*
It is right to increase human welfare when the good done outweighs the harm suffered.		It is never right to make a majority suffer for a minority (1) if the gains are trivial, or (2) if the suffering is substantial.

MINOR PREMISE

To be determined: (1) Will only a minority use the plane?
(2) Will a majority be affected by its noise? (3) Are the gains trivial?
(4) Are the hardships substantial?

It is worth noting that even the determination of the facts will often not be a purely empirical matter. Perhaps what constitutes a minor inconvenience and a grave harm will be clear to both sides; but it may also turn out to be an evaluative matter subject to dispute, in which case the work of justification will begin again. In any case, it is possible for an argument to unfold, even in public debate, largely on the level of moral principles. Most significant debates do contain some sparring over just what principle applies. It is assumed, of course, that the parties to the debate will have sufficient objectivity to acknowledge the legitimacy of the moral principles invoked by their opponents. If not, then either there will be an impasse, or the argument must move back one step as the challengers try to show why their principle ought to be accepted.

"MORAL REASONS"

One final point—we have talked about the process of moral reasoning, but we have not indicated what kind of principle constitutes a *moral* principle. There are two ways to conceive of a moral reason for doing or refraining from doing something. The first is to say that a moral reason must be other-regarding; that is, it must take account of values other than self-interest, and particularly it must acknowledge the worth of other persons. The principle "Do no harm!" is an example of an other-regarding principle. It assumes that because other people and things are of value, they should not be harmed without serious reason. Secondly, a moral reason is often considered to be an impartial reason—that is, a rule any objective and disinterested person would be willing to abide by. The principle of fairness or justice represents the impartial moral rule. Thus, the rule "It is never right to make a majority suffer for the benefit of a minority . . ." derives from a kind of neutral standpoint in which all individuals count equally in the distribution of a society's burdens and benefits.

Together, other-regardingness and impartiality serve to distinguish moral reasons from other kinds of explanations, particularly arguments stemming from utility or welfare. A cost–benefit assessment, for instance, is not a moral argument until it touches on questions of distribution (justice) or basic goods (rights). For example, the argument from social welfare, as we called it, is a nonmoral argument until we begin to examine who the affected population will be and how they will be affected. In the abstract, the construction of the SST might increase the output of goods and services in our society, but in the concrete, it places unfair and serious burdens on a great many people. The social welfare argument is disqualified as a moral argument because it fails to take into account other important moral values. (While, in principle, techniques like cost–benefit analyses might be employed in such a way as to reflect our moral judgments, in practice they do not. The medical costs to people whose hearing is impaired by jet noise in no way reflect the real costs of a hearing deficiency or, on the moral side, the violation of bodily integrity that noise pollution involves.) Thus, for intelligent discussion of policy issues, distinctively moral arguments need to be distinguished from arguments about the economic advantages of scientific programs.

Taken seriously, the concepts presented in this chapter imply that the mission of science instruction in the coming decade must be greatly broadened. Some fundamental curriculum development and experimental teaching need to be done. It is the kind of venture in which many citizens will eventually need to take part so that we will all develop the intellectual skills necessary to engage in the dialogues required for participation in this scientific/technological/human era.

CHAPTER 2

What Research Says About Student–Student Interaction in Science Classrooms

Roger T. Johnson
David W. Johnson

In every science lesson, the science teacher structures the way in which students interact with each other as they pursue their learning goals. Science teachers can structure student learning goals so that students are in a win–lose struggle to see who is best, so that they work independently of their peers, or so that they work in pairs or small groups to complete the assignments and help each other master the assigned material. Or science teachers can structure some mixture of these three basic goal structures. Whether science teachers structure learning situations competitively, individualistically, or cooperatively will determine how students interact. These interaction patterns, in turn, determine instructional outcomes.

In this chapter the relevance of structuring student–student interaction appropriately in science classes is explored. Competitive, individualistic, and cooperative learning situations are defined, and the research comparing the relative effects of these three goal structures is then reviewed. Finally, the specific procedures for conducting cooperative science lessons are discussed.

CURRENT PRESSURES ON SCIENCE TEACHING

A number of factors are presently having a strong influence on science education. The "back-to-basics" emphasis evident in American education for several years has affected science instruction. In the elementary school, "back-to-basics" has often meant reducing the time spent in science instruction or eliminating science instruction altogether. As a result of decreasing achievement scores in science,[1] science teachers have moved toward more "whole-class" instruction, toward more lecture- and textbook-dominated instruction and less laboratory and small-group instruction. The current striving for achievement gains in science has resulted in de-emphasis of desired affective outcomes, such as building positive attitudes toward science. Student attitudes toward science as a subject area are not encouragingly favorable.[2]

Should science be sacrificed in order to improve achievement in reading and math? No. Is there value in striving to increase science achievement if, at the same time, interest in science and science-related careers is decreased? No. We do *not* have to choose between increasing science achievement and building interest in science-related careers. The research on student–student interaction patterns indicates that both cognitive and affective outcomes can be obtained at the same time.

COMPETITIVE, INDIVIDUALISTIC, AND COOPERATIVE INSTRUCTION

The way in which instructional goals are structured controls the nature of student–student interaction which, in turn, controls instructional outcomes.[3] Science teachers can structure learning goals competitively, individualistically, and cooperatively.

Competition among students is caused by negative goal interdependence; students perceive that they can obtain their goal if, and only if, the other students with whom they are competitively linked fail to achieve their goals. Students are instructed to try to work faster and more accurately than their classmates, they are graded using a norm-referenced system, and the winners are rewarded. This "Do better than your classmates" situation occurs, for example, when science students are evaluated on a curve to see who has done the best work on an experiment.

During *individualistic* work by students there is no goal interdependence; students perceive that their goal achievement is unrelated to, and independent from, the goal achievement of other students. Students work on their own, with their own set of materials and at their own pace, without interacting with other students. They are then rewarded on the basis of how their performance compares with preset criteria of excellence. An example of this "Work on your own" situation occurs when each science student is required to complete an experiment without interacting with other students; each student knows that her or his work will be evaluated using a criteria-referenced system so that the achievement of one student has no effect (positive or negative) on the achievement of other students.

Cooperation among students is encouraged by positive goal interdependence; students perceive that they can obtain their goal if, and only if, the other students with whom they are cooperatively linked achieve their goals. Students are instructed to work together to achieve a group goal, are evaluated using a criteria-referenced system, and are rewarded on the basis of the quality of the group's product. An example of this "Sink or swim together" situation occurs when a group of students conducts an experiment and agrees on one set of answers, while ensuring that each group member is able to explain the rationale for the answers.

Recent tradition in schools encourages interpersonal competition in which students are expected to outperform their peers. When a student enters school, there is great concern over whether her or his performance is equal to or better than that of other students in the class. To know more than others is taken as a sign that one is better, more intelligent, superior; and being more knowledgeable is prized. Constantly encouraging students to outperform their peers has had considerable socializing effects, as indicated by the facts that American children are more competitive than children from other countries and that they become more competitive the longer they are in school or the older they become.[4] Not only do most students perceive school as a competitive enterprise,[5] but also Nelson and Kagan[6] conclude on the basis of their research that American students so seldom cooperate spontaneously that it appears that the environment provided for them is barren of experiences that would sensitize them to the possibility of cooperation.

Individualistic instruction, during which students work alone with their own set of materials toward their own learning goal, has been presented as an alternative to competition and implemented widely in the past 10 or 12 years. Yet, it seems to contribute to student loneliness and alienation and to have an adverse effect on socialization and on healthy social and cognitive development.

Although clustering students together to work is not uncommon in science classes, cooperation is the least used of the three goal structures. Cooperation is *not* having students sit close together, each doing her or his own work but talking with one another. Nor does cooperation exist when one student does all the work for the group while three others go along for the ride. Cooperation is *not* having students share materials or equipment before they take a competitive test. Cooperative interdependence means that the students perceive their success to be dependent on the efforts of all the members of their group so that their efforts as a group are evaluated against preset criteria of excellence and all members of the group must master the assigned material. Positive goal interdependence with individual accountability is often not stressed in "group work" or "lab groups," and students end up discussing and working parallel rather than fully collaborating.

Of all subject areas, science, with its emphasis on problem solving, laboratory investigations with concrete materials, and divergent thinking, is especially well suited to having students work together in cooperative learning groups.[7] Science teachers often seat students together to work (even when the teachers are not sure that they want students to interact) due to a short supply of materials or the number of lab tables in the classroom. What have been missing from these often loose and casual clusters of students are specific strategies for how to structure cooperation among students so that each group works effectively, maximizing each student's achievement while at the same time building positive attitudes toward science as a subject area.

RESEARCH ON STUDENT–STUDENT INTERACTION

There have been several hundred studies conducted comparing the relative impact of competitive, individualistic, and cooperative instruction on cognitive and affective outcomes.[8] These studies have covered a wide range of age levels (preschool through adult), subject areas (including science), and tasks (including many that are science-related). It is clear from the studies that working as part of a cooperative group is a very powerful way to learn, and it has a powerful effect on many different learning outcomes. In this section we will focus on a number of instructional outcomes, starting with the cognitive area and moving to affective outcomes of instruction.

Achievement

Science achievement, especially in the physical sciences, appears to be steadily decreasing. A comparison of three national assessments of science indicates that students at ages 9, 13, and 17 knew less in the physical science areas in 1972–1973 than they did in 1969–1970, and even less in 1976–1977 than they did in 1972–1973.[9] While the overall decay of science achievement was somewhat halted for students at ages 9 and 13 between 1973 and 1977, due to a comeback in biology, the 17-year-olds continued their decline in both the physical and the biological areas during those years. A look at SAT scores during these years indicates that the decline in achievement is not only a science problem but also a schoolwide problem; and a close look at juvenile crime, suicide rates, and other societal data indicates that it may be more a reflection of a societal crisis than a school problem.[10] What, then, can be done to improve achievement in science classes?

The research on student–student interaction clearly indicates that structuring students to work cooperatively promotes higher individual achievement than does structuring students to work competitively or individualistically.[11] In order to analyze virtually all the studies that have ever compared the effects of cooperative, competitive, and individualistic

28

learning situations on achievement and productivity, we have conducted a series of meta-analyses on the results of 122 studies. The meta-analyses indicate overwhelmingly that cooperative learning promotes higher achievement than do individualistic and competitive learning (the probability that this conclusion is due to chance is less than 0.00001). The student working at the 50th percentile under competitive and individualistic conditions achieves at approximately the 80th percentile under cooperative conditions. In a series of subanalyses to find possible mediating variables, the superiority of cooperation in promoting achievement is consistent over all age levels, in all subject areas, and for all types of tasks (with the exception of rote decoding and correcting). The higher achievement promoted by cooperation is especially marked in conceptual and problem-solving tasks. In addition, a number of studies either in the science area or dealing with science-related tasks have found that students not only achieve more but also they retain the information longer.[12]

One answer, then, to the question of how to raise achievement in science classes within the existing curriculum is to carefully structure students to work collaboratively much of the time. The discussing, explaining, arguing, teaching, and encouraging each other to learn that are part of a cooperative group have positive effects on individual students' success in learning and in retaining science material and methods.

Science Controversies Among Students

Controversy exists when one student's ideas, information, conclusions, theories, or opinions are incompatible with those of another student and the two seek to reach an agreement.[13] While the field of science thrives on controversy, it is discouraged or even forbidden in a great many science classrooms as being a sign that students are not interacting effectively. In the NSF-sponsored case studies of science classrooms directed by Stake and Easley[14] science teachers are found to be highly concerned that students get along with each other. When students disagree and argue over ideas, conclusions, and opinions, it matters a great deal whether teachers are aware of the potential value of controversy and have the skill to structure a cooperative context for it. The latter is especially important because the evidence indicates that controversy has positive influences on learning only within a cooperative context; students who are competing with each other tend to get "locked in" by the win–lose dynamics, and controversy becomes a negative influence on learning.[15]

When controversy occurs within a cooperative learning group, certain consequences have been identified. Controversy begins, as does all learning, with students categorizing and organizing their present information and experiences so that a conclusion is derived. When the participants realize that others have a different conclusion and that their conclusion is being challenged, the following take place:

1. Students become uncertain about the correctness of their conclusions.

2. In hopes of resolving their uncertainty, students actively search for more information, new experiences, and a more adequate cognitive perspective.

3. In actively representing their position and reasoning to the opposition, students cognitively rehearse their conclusions and rationale.

4. In their search for a more adequate cognitive perspective, students listen to and attempt to understand their opponents' conclusions and rationale.

5. The cognitive rehearsal of their own position and the attempts to understand their opponents' position result in—
   a. Mastery and retention of the material being learned.
   b. An accurate understanding of their opponents' cognitive perspectives.

6. The uncertainty does not fully end when a joint conclusion is reached, and, therefore, there is continuing motivation to learn more about the issue.

7. The process of arguing and coming to a joint conclusion creates positive attitudes toward science and the controversy procedures, interpersonal liking, and positive attitudes among students.

Several studies in the science area have shown that controversy in cooperative learning groups (compared with the absence of controversy in groups, competition, and working alone) results in greater mastery and retention of the subject matter, higher quality problem solving, the transition to higher levels of cognitive and moral reasoning, shifts in judgment, more and higher quality ideas, and more accurate perspective-taking.[16] The more skillful students become in managing controversy, the more valuable it becomes as a procedure to enhance the learning situation. These repeated "friendly" excursions into disequilibrium by individuals in cooperative groups is an important aspect of science education.

Attitudes Toward Science

There is ample evidence that students' attitudes toward science are not positive. The National Assessment of Science conducted in 1976–1977 included a number of items designed to measure students' attitudes. At first glance, the data seem to indicate that 9-year-olds are reasonably positive about science. Almost two-thirds of the 9-year-olds indicate that they are happy with science and over four-fifths indicate that they are interested in science. Only about half of the 9-year-olds, however, indicate that they feel excited or successful in science, and they rank science as one of their least favorite subjects in comparison to others such as math and English.[17] By age 13, one-third of the students indicate that they do not want to take any more science than they have to, and another one-third are not sure. Less than half of the 13-year-old students indicate interest in a science-related career. By age 17, attitudes have decayed further. The overall picture is one of steady decay of attitudes toward science from elementary school through junior high and high school. The Minnesota School Affect Assessment, which has been administered widely in Minnesota and several other states, shows not only a drop in positive attitudes toward science but also a drop in attitude scores from fall to spring at each of these school levels.[18]

Structuring the interaction among students appropriately in science classes is a powerful strategy to promote more positive attitudes toward science. Several studies have shown that students' attitudes toward the subject area and its instructional activities, as well as their continuing motivation to learn more about the subject, are more positive and higher under cooperative instruction than under competitive and individualistic instruction.[19] If science educators wish to promote more positive attitudes toward science, enjoyment of science activities, and continuing motivation to learn more about science, a cooperative learning structure is to be preferred over competitive and individualistic ones.

The positive attitudes toward subject areas found consistently in cooperative learning situations are manifested in several ways. There is greater involvement in instructional activities when cooperation is used. The more cooperative the students' attitudes, the more they say that they express their ideas and feelings in large and small classes and listen to the teacher; competitive and individualistic attitudes are unrelated to such indices of involvement in instructional activities.[20] These and other studies indicate that the more favorable the students' attitudes toward cooperation, the more they believe that teachers, teacher aides, counselors, and principals are important and positive; that teachers care about and want to increase students' learning; that teachers like and accept students as individuals; and that teachers and principals want to be friends with students.[21] It is clear that teachers can encourage more involvement in instructional activities and build more positive attitudes toward science and science teachers by placing students in collaborative relationships instead of having students compete with one another or work alone.

Attitudes of Students Toward Each Other in Science Classes

One of the major challenges that schools as well as science classes will be facing in the 1980's is how to constructively deal with heterogeneity of students in the classroom. Integration of different ethnic groups will remain an issue, and educators will begin to focus less on the percentages and quotas within a school and more on the quality of interactions among ethnic groups in the classroom. More handicapped students will be placed

31

in regular classrooms for more of the time. Concern is increasing for ways to capture and maintain the interest of females in science as a subject and a possible career. The NAEP analysis of the three national assessments (1969–1970, 1972–1973, and 1976–1977) reveals that black students, female students, and students from disadvantaged–urban communities consistently perform below the national level in achievement at each age level.[22] Other national assessment data indicate that females are already showing less interest in science at age 9 and that the gap between the interest of males and females in science widens at ages 13 and 17.[23] Science is not only losing the interest of many students, but also it is losing the interest and involvement of specific subgroups of students, primarily minorities and females. Minorities, handicapped students, females, and other students who add to the heterogeneity of the science classroom need to be socially integrated in a way that facilitates their learning and their interest in science and science-related careers.

There is considerable evidence that cooperative learning experiences, as compared with competitive and individualistic ones, result in more positive interpersonal relationships among students characterized by mutual liking, positive attitudes toward each other, mutual concern, friendliness, attentiveness, feelings of obligation to other students, and a desire to win the respect of other students.[24] Cooperative learning experiences build these positive relationships among students from different ethnic groups, different social classes, and the opposite sex, as well as among classmates who are intellectually or physically handicapped and nonhandicapped. Cooperative relationships among heterogeneous groups of students tend to produce acceptance of differences, promote a view that differences are an enriching resource, and enhance the exploration of different perspectives. Competitive and individualistic instruction do *not* tend to build acceptance of differences (or friendships in general), but rather they encourage students to associate with others who are perceived to be similar and to reject students who are perceived to be different. The positive attitudes among students, as well as the peer support and encouragement for learning that are nurtured in heterogeneous, cooperative groups, tend to counteract negative pressures that work against learning science. Students not only learn more, but also they learn to accept one another as individuals and to like each other regardless of differences.[25]

Self-Esteem

Many students feel inadequate in the area of science. A feeling that "Science is more than I can handle" is reflected in the responses of high school students to a nationwide poll conducted by the Purdue Opinion Panel.[26] This survey has found that 35 percent of high school students believe that it is necessary to be a genius to be a good scientist and that 30 percent believe that one cannot raise a normal family and become a scientist. One aspect of teaching science, therefore, may be to promote a sense

32

of adequacy and self-confidence in students with regard to their ability in the science area. Science teachers also need to be concerned about the levels of self-esteem of their students for a variety of other reasons including their students' psychological health, achievement in science classes, general happiness, and selection of a science-related career.

How the student–student interaction in science classes is structured does affect students' self-esteem. Correlational studies indicate that cooperativeness is positively related to self-esteem in students throughout elementary, junior high, and high school in urban, rural, and suburban settings; that competitiveness is generally unrelated to self-esteem; and that individualistic attitudes tend to be related to feelings of worthlessness and self-rejection.[27] Further analyses reveal that positive attitudes toward cooperation are related to basic self-acceptance, which is relatively stable, while positive attitudes toward competition are related to conditional self-acceptance, which tends to rise and fall with performance. There is also evidence that cooperative learning experiences, as compared with individualistic ones, result in higher self-esteem.[28] To the extent that science instruction is structured cooperatively and encourages positive attitudes toward cooperation, higher levels of self-esteem and more basic self-acceptance will be encouraged.

Psychological Health and Social Skills

The ability to build and maintain cooperative relationships is vital to most science-related careers and is often cited as a primary manifestation of psychological health. In a study comparing the attitudes of high school seniors toward cooperation, competition, and individualism with their responses on the Minnesota Multiphasic Personality Inventory,[29] cooperative attitudes are significantly and negatively correlated with psychological pathology (9 of the 10 scales), as are competitive attitudes (7 of the 10 scales). Attitudes toward individualism are significantly and positively related to 9 of the 10 psychological pathology scales. These findings indicate that an emphasis on cooperative involvement, and appropriate competition, with other people *may* promote psychological health and well-being, while social isolation *may* promote psychological illness. Science instruction can make a contribution to the well-being of students by structuring appropriate interaction among students and by eliminating the extensive use of individualistic modes of teaching.

Producing students who are well versed in science but unable to work collaboratively with others will be of little use to society. The cooperative skills needed to maintain career, family, and community relationships, as well as friendships, are basic to every individual. Cooperative learning experiences, as compared with competitive and individualistic ones, not only promote the development of the basic interpersonal skills such as communication, leadership, and trust-building, but also they promote the interpersonal skills needed to engage in the problem solving and inquiry that are encouraged in many science curriculums. Cooperative experiences promote increased competence in perspective-taking, mutual influence, coordination of efforts, and divergent thinking.[30] The cooperative skills students develop as they work in cooperatively structured science classes are as important to the students as science skills. Science knowledge and skills are of little use if students cannot apply them in cooperative interaction with other people.

Conclusions

The research on student–student interaction patterns is extensive, and the results are impressive in terms of the benefits of cooperative learning. A couple of perspectives on this research, however, may help to clarify its value.

Structuring cooperative learning in science classes is not a "magic wand" that will suddenly solve all problems and raise all individuals to the highest levels of achievement with positive feelings about science as a subject and a possible career. Rather, the research indicates that we will have a *better chance* of accomplishing our goals as science educators if we structure learning situations cooperatively than we will if we structure them competitively or individualistically.

A second perspective is that a carefully orchestrated combination of all three goal structures would probably be the most desirable method to meet the needs of teachers and students. The variety that a mixture of cooperative, competitive, and individualistic learning experiences would provide during a day or a week would allow students to learn not only how to collaborate skillfully but also how to work autonomously to complete an individual task and how to enjoy competition, win or lose. All three student–student interaction patterns have value. It is cooperation, however, that is the most powerful and most useful.

THE SCIENCE TEACHER'S ROLE IN STRUCTURING COOPERATION

With any good idea, it is always necessary to gather your own data with your own students and in your own situation. A description of how to structure the different interaction patterns appropriately is found in *Learning Together and Alone.*[31] You may wish to try out each student–student interaction pattern with your students for a week or so and discuss with the students the advantages of each. Care must be taken in structuring cooperation to make sure that students realize that they are in a "sink or swim together" relationship and that each group member is accountable for learning the assigned material. Do not be surprised if at first several of your students do not function well in groups; many students are not very skilled at working collaboratively. Some social-skill teaching may be needed before students are fully capable of working cooperatively.

A brief overview of the strategies for structuring a cooperative lesson should be useful. There are four major jobs for the science teacher in setting up a cooperative learning situation: (1) a set of initial decisions must be made (e.g., group size, distribution of materials), (2) the cooperative structure must be explained to the students, (3) student behavior must be monitored, and (4) cooperative skills must be taught. The following guidelines are not a formula, but rather a model that many science teachers have found helpful.

Assigning Groups

The first step for a science teacher in structuring a cooperative learning activity is to determine an appropriate group size and assign the students to groups. You need enough members in each group to stimulate each other's thinking, but not enough to allow one or more students not to participate. Start with small groups and work your way up to larger groups as students become more skillful in collaborating. One way heterogeneous groups are formed is by having students count off randomly. The "ones" become a group, the "twos" become a group, and so forth. The intent is to form heterogeneous groups in which students have different backgrounds, perspectives, and skills. Heterogeneous groups are potentially the most powerful in problem-solving situations. Sometimes students may ask if they have to work with the group they have been assigned to; they can be told that eventually they will work with everyone in the class and that this is the group they will be working with today.

Arranging the Room and Distributing Materials

The second step is to arrange the room and the science materials to promote collaborative interaction. Each group sits close together, separated as far as possible from the other groups. The materials and apparatus are set on a centrally located table. One set of materials may be assigned to each group.

Assigning Cooperative Learning Goals

The third step is to assign the science task, describe the cooperative structure of the learning goals, and communicate the criteria for evaluation. Assign the task in a clear and specific way. One way to set the cooperative goal structure is to ask each group to submit at the end of the class period *one* report describing the members' best opinion and the group's rationale for the decision. Group members are told to sign the report only if they agree with the answer and can explain the rationale. The students are told that they should share ideas, listen to each other's ideas, participate in the testing process, ask other group members to verify the results, and double-check the test results and information with other groups when it seems necessary. The evaluation criteria may be as follows: if the group report is 100 percent correct, the group has done an excellent job; if the report is partially correct, the group has done an acceptable job; and if the report is way off, the group is sent back to the drawing board.

Observing—Formally and Informally

The fourth step is to observe the groups of students. Assigning students to groups and instructing them to work cooperatively will not mean that they will or can do so. Teachers can use a formal observation sheet to check specific aspects of cooperative interaction—e.g., actively participating, actively listening to other students' comments, presenting logical answers to the group, and summarizing data and the group's conclusions. After teachers observe and model how to share observations without making judgments about students' actions, students can be given the opportunity to use the formal observation sheet. Observing a group is probably the best way for students to learn cooperative skills—they concentrate on the presence and absence of the skills and see them used or misused as the group learns science.

As teachers move around the room, they can also make informal notes. When observing student–student interaction, teachers can jot down a few notes to expand on later as they reflect over the observational and achievement data and draw conclusions about progress in learning science and collaborative skills.

Intervening To Teach Social Skills

The fifth step in conducting a cooperative science lesson is to intervene in groups that have difficulties in working together. A group, for example, may be leaving out one member. The teacher may stop the group and point out that because not everyone is being included, the group is losing resources and will have difficulty getting everyone to sign the report at the end of the class period. In response to the teacher's observation, the group may decide to check every few minutes to make sure that everyone is participating and understands what the group is doing. The group then proceeds with the science task. The teacher may wish to watch for a moment or two and then move on to observe another group.

The basic cooperative skills students need to master are trust-building, communication, leadership, and conflict resolution.[32] Collaborative skills must be defined so that students will understand how to behave cooperatively. Some groups may have problems integrating the cooperative skills into their efforts to complete the assigned task. Teachers must be able to say, "Put away the task for a few minutes. We have a cooperative-skills problem to solve." The teacher then stays with the group as the group solves its problem.

Evaluating Students' Work

The final step is to evaluate the quality and quantity of the students' learning. The teacher collects the group reports and evaluates them according to the preset criteria. Students may evaluate the functioning of their groups by spending the last 10 minutes of the period discussing how well they worked together and each member's contributions. The teacher may offer some overall observations and have the groups summarize their ideas for working together more effectively to share with the entire class.

SUMMARY

In order to simultaneously maximize achievement in science classes and promote positive attitudes toward science and science-related careers, science teachers need to use competitive, individualistic, and cooperative learning situations appropriately. In light of the research findings, there is little doubt that science teachers who predominantly use the cooperative learning structure will be able to have a powerful and positive effect on their students' achievement and attitudes. This addition to the teacher's repertoire does not mean establishing a new curriculum or spending any money. It only takes a few minutes to make clear to students the kind of student–student interaction expected, with some additional time needed in the beginning to teach students the appropriate interaction skills. All teachers are interested in the achievement gains promised by the research. Teachers may be even more interested in promoting positive attitudes toward science and increased social skills in their students. The initial effort by teachers and the time needed to structure student–student inter-action carefully are well worth it. It would be exciting to see the gap dis-appear between the research findings and traditional science classroom practice, so that science students would reply to the question, "How do you see school?" by saying:

> School is a place where we work together to learn. We share our ideas, argue our points of view unless logically persuaded to change our minds, and help each other find the most appropriate answers. Occasionally we have a fun competition or work on our own, but most of the time we learn together.

CHAPTER 3

Linking Teaching Behaviors and Student Behaviors in Science

James R. Okey
David P. Butts

The problem *seems* straightforward. If we find out what teaching behaviors are associated with high pupil achievement in science, teachers can use the behaviors and students will learn. Educators, parents, taxpayers, and even students should be happy with that. But the problem isn't that easy. Some critics of education take the stance that teaching is an art. They feel that attempting to make it a science by establishing cause-and-effect links between what teachers do and what students learn is nearly fruitless. For years this view of teaching as an art was hard to refute because compelling evidence to the contrary was hard to find. However, in the last 10 years the picture has changed considerably. Increasingly, studies are reported that demonstrate that it is possible to identify effective teaching behaviors.

The purposes of this chapter are to describe how strategies of research on teaching behaviors have changed over the last 30 years and to present some of the more promising results of recent research efforts that may serve as a guide to science teachers.

RESEARCH TECHNIQUES APPLIED TO TEACHER AND STUDENT BEHAVIORS

Studying Teacher Characteristics and the Teaching Model

If you examine reports of teaching research from 20 years ago, you will find that many of them concentrate on only half of the teaching–learning equation.[1] Most of the interest is focused on teachers—i.e., the characteristics of teachers considered important in teaching. The researchers were seeking a description of the master teacher—because if a description of one could be formulated, then the task of novice teachers (or experienced ones) was to pattern or model their behavior after the description of the master teacher. Young teachers were encouraged to "model the master teacher" if they wished to be more effective instructors.

Some of the problems with this master teacher approach to identifying teaching behaviors are addressed by Stolurow[2] who thinks that the focus of the research is misdirected. According to Stolurow, we should not be attempting to "model the master teacher" simply because human characteristics are so numerous and diverse that their study is fruitless. Instead, we should be trying to "master the teaching model"—that is, to understand the components of instruction that are essential to learning. This type of thinking lead research on teaching away from concerns about human characteristics such as sincerity, empathy, or friendliness, and toward an examination of instruction that focuses on such factors as clarity of outcomes, sequences of tasks, provision of appropriate practice, and feedback opportunities.

Research that followed this teaching-model conception sought components of instruction that might directly influence student outcomes. Usually called *process–product research*, this approach has attempted to equate what teachers do (processes) with what students accomplish (products). Figure 1 shows the two items of major concern in process–product research. The teacher behaviors examined include questioning strategies, clarity of explanations, and task orientation, as well as such factors as teacher warmth, directness, and enthusiasm. Thus, the teacher behaviors under investigation included both teacher characteristics and teacher actions. The second box in the model, student outcomes, refers primarily to student achievement, although it might also refer to attitude gains or to such outcomes as school attendance.

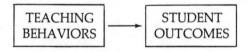

FIGURE 1
THE PROCESS-PRODUCT TEACHING MODEL

An influential paper by Rosenshine and Furst[3] has given prominent attention to process–product research and has attempted to synthesize the findings for 11 teacher behavior variables. Although their paper gave major impetus to process–product research, gains in knowledge about teacher behaviors and student outcomes were hard won.

Why does the process–product model from Figure 1 fail to shed much light on the relationship between teacher behaviors and student outcomes? An explanation has been put forth by Medley, Soar, and Soar.[4] They contend that an important part of the model is missing (see Figure 2). Teachers don't directly influence pupil achievement, they reasoned; therefore, searching for direct cause-and-effect links between teaching behaviors and student outcomes is likely to yield little information. Instead, they argue that teacher behaviors influence pupil study behaviors. In other words, a science teacher has direct control not over what the student learns but over what, how, and how long the student studies science. What *does* directly influence the student's science achievement? It is the student's study behaviors. So the contribution of Medley, Soar, and Soar has been to interest researchers in this intermediate step in the teaching–learning model.

FIGURE 2
THE PROCESS-PRODUCT MODEL WITH AN INTERMEDIATE STEP [5]

Measuring Teaching Behaviors and Student Study Behaviors

In order to study what teachers and students do in science classrooms, measuring devices are needed. The measures are descriptions of teacher and pupil behaviors that guide observers in seeing what goes on in the classroom.

The basic characteristics that any measurement device must have are validity and reliability. This means that the measure should be an accurate or true indicator of the behaviors in question and that it should consistently provide evidence about them. Literally hundreds of such measuring instruments have been developed over the years. The instruments vary widely:

1. Some are limited in scope (e.g., only teacher questioning behavior), while others are broad (e.g., all verbal interaction).

2. Some result in numerous pieces of information (e.g., behavior is coded every three seconds), and others provide more sparse descriptions of classroom activities.

3. Some are limited to directly observable behaviors (e.g., how often a teacher smiles), while others require higher inference judgments by observers (e.g., appropriateness of objectives for students).

4. Some deal only with teacher behaviors while others deal with both teacher and student activities.

Two available compilations of classroom observation instruments are *Mirrors for Behavior* [6] and *Elementary Classroom Instruction.* [7] Well over 100 instruments for observing in classrooms are available in these two volumes.

A classroom observation instrument set developed recently in the state of Georgia provides a comprehensive analysis of a teacher's activities. [8] It includes observation of a teacher's plans and materials, classroom procedures, interpersonal skills, and professional activities.

Two observation instruments designed especially for science classrooms are available. One of these, the Teaching Strategies Observation Differential (TSOD), allows an observer to code 10 different types of classroom environments being used by the teacher. The environments range from structured lecturing to student-designed investigations. [9] The second instrument, the Data Processing Observation Guide (DPOG), focuses on science classrooms in which the teacher has students collecting, displaying, and interpreting data. [10]

Student study behaviors are observed in a variety of ways, but the most common is to assess the time during which they are intellectually engaged with the subject matter. Students are intellectually engaged when they are involved in such activities as listening, taking notes, talking with a fellow student or teacher, or performing an activity associated with the topic to be studied. Intellectually engaged time is usually referred to as *on-task time.* Capie, Dillashaw, and Okey [11] describe a procedure for measuring the on-task behavior of students that consists of observing a student for three or four seconds and judging whether he or she is involved in the topic. Using this procedure, an entire class of 30 students can be observed in less than three minutes. Multiple observations of each student are usually made. The coded information allows statements (usually expressed in percentages of the total time observed) about the on-task behavior of individual students or of the entire class.

Student outcomes (the third box in Figure 2) are usually measured in terms of cognitive achievement. Teacher-made or standardized tests are the most common measures used. Attitudes, perceptions, and similar affective outcomes may also be measured. Occasionally, because less conventional outcomes are of interest, special tests or measurement devices need to be located or developed. Examples of these outcomes would be amount of student talk, level of cognitive development, and degree of cooperation among students. *Diagnosing Classroom Learning Environments* by Fox, Luszki, and Schmuck [12] includes a variety of instruments for measuring social, emotional, and perceptual outcomes in classrooms.

Correlational and Experimental Studies

Two kinds of investigations are possible when studying the teaching behaviors of science teachers. The distinction between the two can be understood by considering an example. Suppose a researcher wants to study the amount of time that a science teacher waits after a student begins to respond before interjecting a comment or follow-up question. The speculation is that students give more complete answers when their teacher gives them more time to respond. One way to study this "wait time" is to observe dozens of classrooms and hope that a variety of teachers who wait very little and teachers who wait a long time will be found. An alternative approach to studying the question is to select a smaller number of teachers (perhaps a dozen) and instruct them to use specified wait times. Some teachers are instructed to use short waits, others wait times of intermediate length, and still others long wait times. There are similarities in the two approaches; for example, teachers with differing wait times are found in each. But there are important differences, too. In the first instance, the differences in wait time occur naturally, and in the second they are deliberately manipulated by the researcher.

The first approach allows the classroom researcher to determine if the two variables—teacher wait time and completeness of student response—are related. Does one increase as the other one does? But the first approach does not allow the researcher to conclude that one variable influences or causes the other. This first type of study is referred to as a *correlation study* because it relates two variables.

The second type of study described above, in which the wait time of teachers is deliberately changed, is called an *experimental study*. The researcher manipulates one variable (teacher wait time) and checks to see if the second variable (completeness of student response) is changed. This type of study with deliberately manipulated variables allows cause-and-effect conclusions to be drawn.

Both correlational and experimental studies are important. Even though we want cause-and-effect conclusions about science teaching behaviors and student outcomes, it may be more feasible or cheaper to do correlational studies. The caution we need to keep in mind, however, is this: the fact that two variables are correlated does not necessarily mean that one causes the other.

The remainder of this chapter describes a variety of studies about teaching behavior and student outcomes. Some of them are correlational studies and some are experimental. Sometimes the teaching behaviors are obvious actions (e.g., questioning behavior), and sometimes they are more elusive characteristics (e.g., enthusiasm or rapport). The studies also show great diversity in the types of outcome. In some, the study behavior of students (e.g., on-task time) is being measured, and in others academic accomplishment (e.g., chemistry achievement) is the outcome of interest. What holds the studies together is that (1) they are conducted primarily in science settings, and (2) they identify some of the relationships (both correlational and experimental) among teaching behaviors, student study behaviors, and student achievement.

SPECIFIC LINKS AMONG TEACHING BEHAVIORS, STUDENT STUDY BEHAVIORS, AND STUDENT ACHIEVEMENT

From the research literature, the science teacher behaviors that make a difference in what students do or know can be grouped into six categories:

1. *Subject matter competence* of the teacher

2. *Structure* of the lesson

3. *Physical organization* of the classroom

4. *Motivation* of the students

5. *Rapport* between the students and teachers

6. *Response opportunity* for the students.

Subject Matter Competence

Subject matter competence is the category not often seen in descriptions of teaching behaviors; yet, it is implied in several of them. Gage [13] has described 11 teaching activities, 5 of which are related to subject matter competence: (1) explaining activities, (2) questioning activities, (3) demonstrating activities, (4) assignment-making activities, and (5) curriculum-planning activities. Sorenson and Husek [14] have analyzed the teacher's role along six dimensions, one of which, information-giver, bears directly on the subject matter competence of the teacher. Hord [15] has also included this category in the scheme being used to study and evaluate teaching behaviors. Tollefson [16] has identified 20 items agreed upon by 1,643 high school students as being characteristic of effective teachers. Of these, two deal directly with subject matter competence: "knows his subject" and "doesn't rely completely on the textbook." Teachers are not usually effective in helping students learn what they (the teachers) do not understand. Rosenshine and Furst [17] report that "teacher clarity as assessed by rating scales by students or observers yielded significant results in all seven studies in which the variable was used." Clarity as used by Rosenshine and Furst includes such ideas as the clarity of the presentation, whether the points that the teacher makes are clear and easy to understand, whether the teacher is able to explain concepts clearly, and whether the teacher has the facility and background to answer questions intelligently. After working with 17 biology teachers and 100 biology students, Hart and Towes [18] have found that the science teachers who understand the BSCS framework have students who have better understanding of the principles and concepts in BSCS. Lamb and others [19] find that as teachers become better acquainted with the content of the lesson, they present ideas better and their students achieve more.

Clearly, teachers cannot keep students on-task if they do not understand the task and are not familiar enough with it to effectively guide or direct students' learning.

Thus, the science teacher who wants to enhance the learning time of his or her students needs to do the following:

1. Understand the science concepts to be presented.

2. Fit explanations to the students' levels of understanding of the concept.

3. Provide a variety of examples for each concept.

Structure

Structure includes all those strategies employed by the teacher to design, present, and evaluate instruction. In this category are included such topics as how to get going and keep going, are students or teachers in control, and is the task clear.[20] Most evaluations of this category look at the direct or indirect nature of the teacher's interaction with the students and at the student-centered or teacher-centered nature of the instruction.[21] The proponents of "discovery methods" in science teaching during the 1960's looked at the dimension of guidance (teacher-centered direction) as an indication of the degree of openness in the teaching of inquiry.[22] Schwab[23] has provided a scale of three levels in teaching inquiry, the lowest consisting primarily of teacher direction and the highest consisting entirely of student direction and initiative. In Tollefson's study of the 20 characteristics of effective teachers, 7 are concerned with structure in terms of organization, presentation, student involvement in planning, and reasonable assignments and expectations.[24] Rosenshine and Furst[25] include aspects of this category as organization or flexibility in the classroom structure. But how can teachers develop a classroom structure that will enhance students' time with "academic" activities and their learning?

First, teachers can enhance the effectiveness of the learning environment. Piper[26] describes teacher management behaviors that can improve the maintenance of a high-engaged-time learning environment. These were first described by Kounin[27] as "withitness" and "smoothness and momentum." Campbell[28] has found that junior high science teachers can enhance learning if they are flexible in adjusting their approach to a class to meet the needs of the students. He also finds that high ability classes seem to "lift" the cognitive level of the dialogue in the science classroom. Medley[29] reports that teaching behaviors that cause less deviant, disruptive student behavior were more effective in classrooms in which there was a higher frequency of deviant and disruptive behavior. Medley also finds that less time spent on classroom management is more effective than more time. Shymansky and others[30] note that students who think they can be self-reliant will tend to depend less on teachers for directions. Helping students to understand the management routine or agenda of the science classroom is one important way to increase their learning time.

A second way science teachers can enhance the structure of the learning environment is to focus on how student time is being used. Medley [31] reports that more class time spent on task-related "academic" activities results in greater classroom effectiveness. Rosenshine and Furst [32] seem to describe this as teacher task-oriented or businesslike behavior—i.e., the teacher stimulates student thinking or acquisition of information and is more concerned about student learning than about student enjoyment. Capie, Dillashaw, and Okey [33] have developed an assessment scheme whereby science teachers can check on-task behaviors in their own classrooms. (Their assessment scheme sometimes shows a surprising lack of difference between the way teachers interact in the various sections of science that they teach.)

When Boulanger [34] had junior high science students describe their teachers, he found that students who had hands-on instruction with direct teacher involvement achieved more than those who had more open, non-teacher-directed, hands-on experiences. McDuffie and Beehler [35] conclude that enthusiasm and good work habits are essential for achievement in junior high science classes. Penick and Shymansky [36] have found that in teacher-centered instruction, students spend a significantly greater amount of time following the science teacher's directions than they do in student-centered instruction. They also find that the amount of student on-task time is about the same in both instructional patterns. In describing a special teaching program for working with juvenile delinquents, Test and Heward [37] find that the amount students achieve is proportional to the amount of time they have available to learn. Koran and others [38] also note that low-ability students seem to benefit more from highly structured science lessons—a finding similar to that of Berliner. [39]

A third way teachers can enhance student learning time and achievement outcomes relates to the methods of instruction they use. Raven and Cole [40] have studied the impact of science instruction that requires students to form mental models—and have found that such instruction enhances the students' acquisition of higher-cognitive-level science concepts. Martin [41] finds that students who have behaviorally stated objectives achieve more than others studying similar science content without such objectives.

It is always reassuring when research supports natural wisdom, as is the case here; in summary, the science teacher who wants to enhance student learning time and achievement should do the following:

1. Manage the transitions between learning activities with minimum time and disruption.
2. Maintain a recognized focus for discussion or learning activities.
3. Match the learning activity level to student needs and aptitudes.
4. Minimize deviate and disruptive student behavior.

5. Reduce the time in class needed for management through clearer teacher directions and definite established routines.

6. Increase the amount of time students can be on-task or learning academic content by being businesslike or increasing one's own on-task behavior.

7. Provide students with explicit directions and expectations — increasing the structure when either the task or the students' needs require such a shift.

Physical Organization

Physical organization is a category describing the "background activities" of the teacher — e.g., organizing both student tasks and the classroom environment, managing materials, employing useful items and equipment, and maintaining the focus of attention.[42] This category is implied in 3 of Gage's 11 teacher activities (maintaining order, housekeeping, and recordkeeping)[43] and is stated as one of Sorenson and Husek's six dimensions of the teacher's role, "disciplinarian."[44] Studies of differential teacher interaction with low- and high-ability groups have pointed to poor management and corresponding "nonproductive confusion" as factors correlated with nonachievement.[45]

How the teacher engineers the use of student time through grouping is a frequently explored teacher behavior. Medley[46] summarizes findings in the literature as supporting more effective teaching when students spend less time working in small groups or independently or on seatwork; large groups or smaller groups working cooperatively on well-supervised student worksheets produce more on-task student learning time. Rosenshine[47] also finds that the total time spent in school learning is positively related to achievement if that time is spent on academic activities and negatively related if it is spent on noncurriculum activities. It is interesting to note the absence of studies in science education that focus directly on the strategies a teacher can use in management of either the classroom or time and their relationship to student on-task learning time and learning outcomes. However, some studies have focused on the organizational aspect of the classroom. Gabel and Herron[48] have found that students in ISCS who are taught in a self-paced format without teacher-imposed deadlines do better than those taught in this format with teacher-imposed deadlines. However, Rice and Linn[49] have studied the contrast in on-task learning time and science learning outcomes between students who have free choices in their learning environment and those given directed instruction in that same environment. They have found that students in the directed instruction are more on-task and learn more science. This is in contrast to an earlier study by Linn and others[50] in which students in a free-choice environment achieved more. Rosenshine[51] has reported that in the studies he has reviewed, lowest achievement is associated with students who work independently without supervision and highest achievement with students who work in groups with close supervision.

Thus, a science teacher who wants to enhance the students' learning time and achievement outcomes should utilize these specific strategies related to the classroom's physical organization and time structure:

1. Keep nonproductive confusion to a minimum.

2. Use a variety of resource materials, student activities, and instructional modes.

3. Have students work in large groups, or, if they must work in small groups, cooperatively engage them in the task with close supervision.

4. Reduce the time spent on noncurriculum activities and enhance the time spent on "academic" or objective-related learning.

5. Reduce the amount of unsupervised free choice and increase the amount of time students are working under close supervision.

Motivation

Motivation is as elusive a dimension in teacher activity as charisma is in teacher personality. The motivation of students by a teacher involves a number of considerations. The readiness of students for a given experience is essential and must be determined before expecting them to become involved in a learning task. The science teacher must be able to create an inviting learning situation in terms of organization, assignments, materials, and personal enthusiasm. Interest and involvement on the part of the students are the prime teacher objectives in this category, while sustained interest and achievement of independent learning are the long-term goals of motivation.[52] The learning style of each student is a crucial concern in the motivation category of teaching behavior. Is the student extrinsically or intrinsically motivated, and what can be done to shift the emphasis from the former to the latter?[53] The term *motivation* is repeatedly applied to teachers in specific situations, to high-ability students with personal motivation, and to low-ability students whose lack of motivation stems from repeated failure in the school environment.[54] While this variable does not seem to be found in the science education research literature, Rosenshine and Furst[55] refer to student motivation as a result of teacher enthusiasm. Student ratings of how involved, excited, and interested teachers are in the subject matter are positively associated with learning outcomes.

Thus, a teacher who wants to enhance the motivation level of the classroom as a way of nurturing on-task learning time and achievement outcomes should do the following:

1. Be enthusiastic about what is being taught.

2. Aim instruction at the readiness level of the students.

3. Be interested in the subject matter.

Rapport

The term *rapport* connotes a relationship among teacher and students much like that involved in the achievement of motivation. Is the teacher able to establish two-way communication with the students? Two-way communication involves the expression of feelings—whether enthusiasm or antagonism—and the determination of "who controls what" in the class and the ground rules regarding reward and punishment.[56] The establishment of an effective classroom learning climate is dependent upon the interpersonal relationships among pupils, between teacher and pupil, and between teacher and pupil groups.[57] Low-achieving groups have been described by observers as having "schizophrenic natures"—"half-starved wolves" and "docile lambs" when being taught by different teachers. The teacher of the "docile lambs" considers this level to be a challenge and confesses "a considerable affection" for these students.[58] The category of *rapport* is implied in 2 of Sorenson and Husek's 6 dimensions of the teacher's role, "advisor" and "counselor,"[59] and in 2 of Gage's 11 teacher activities, "guidance" and "mental hygiene."[60] In Tollefson's study, students named seven characteristics related to rapport in their list of 20 characteristics of an effective teacher, employing such descriptors as "available to students," "shows consideration," and "is friendly and enjoys students."[61]

Rosenshine and Furst[62] refer to rapport in their discussion of criticism. Teachers need to let students know when they are on target or correct—or off target or in error (but this need not be done in a critical, I-don't-accept-you mode). They report finding a negative correlation between teacher criticism and student learning in 17 studies. Edwards[63] reports on a strategy that has enhanced the science student's involvement and outcomes. He finds reinforcement from peers to be a powerful alternative to teacher reinforcement. In a later study, Edwards and Surma[64] report that when a science teacher reinforces a student's idea, the student exhibits enhanced inquiry behavior. Rosenshine[65] supports this notion in his summary of four studies in which corrective feedback, even when negative, helps students. He also finds that teacher criticism is consistently negatively associated with student behavior and outcomes.

So a teacher who wants to nurture positive student behavior and learning outcomes through improved rapport should—

1. Maintain true two-way communication with students.

2. Permit students to express feelings within predetermined boundaries.

3. Be available to students as one who listens and accepts them,

even when providing corrective feedback as to unacceptable behaviors.

4. Refrain from criticism.
5. Encourage reinforcement from peers as well as from oneself.
6. Provide corrective feedback when needed.
7. Provide praise and positive motivation.

Response Opportunity

Response opportunity is a category of teacher behavior that has been receiving more attention in recent research. It focuses attention on who is doing the talking how much of the time, on question asking, on keeping the discussion on target, on the level of questions and responses, on the use of student responses, and on maintaining and supporting response opportunities for students.[66] Response opportunity is a factor that has been included, whether stated or implied, in most current studies of teacher–student interaction.[67] Teacher questioning strategies and student responses are receiving attention in specific studies.[68] One recent study investigates the relationships among question level, response level, and lapse time, and analyzes the effects of the teacher's perception on the individual's opportunity to respond.[69]

Rosenshine and Furst's review includes three kinds of teaching behaviors related to this category—businesslike orientation, unidirectness, and multiple level of questions.[70] Their summary suggests that "teachers get what they teach for." Being businesslike or task oriented enhances the focus or direction of the classroom dialogue. The teacher's actions in using student ideas by acknowledging them, modifying the ideas when needed, applying the responses to a new idea, and comparing one student response to another are related to achievement. Medley[71] notes that teachers who use fewer rebukes have more effective learning environments. Rebukes to students decrease the likelihood that they will respond in the classroom. Rosenshine and Furst[72] also note that better learning environments are characterized by teachers who have an explicit beginning and ending for a learning event.

In maintaining these learning events or discussions, DeTure,[73] in an extension of the original research reported by Rowe,[74] finds that a longer teacher wait-time results in more student input to the science discussion. Rice[75] also finds that a longer science teacher wait-time results in greater variety in student thought. Tobin[76] finds that longer teacher wait-time results in greater student achievement in science.

Providing feedback to students may encourage their continued response in the classroom. Rosenshine[77] reports in his review of several studies that the impact of adult feedback (e.g., praise) is neither consistently good nor counterproductive. However, DeBoer[78] finds that frequent testing of students with feedback from the teacher is correlated with greater science achievement. Burrows and Okey[79] also show that the use of diagnostic tests and remediation enhances students' science achievement. Yeany and Capie[80] have noted similar results among college biology students. Pouler and Wright[81] provide evidence that direct instruction and feedback on student responses are more efficient than either indirect instruction or the withholding of feedback in helping students acquire inquiry behavior.

The methods of instruction a teacher uses have been shown to be closely related to student on-task time and learning. Medley[82] notes that effective learning contexts are those in which teachers use mostly low-level questions and few higher-level questions, and in which teachers are less likely to amplify a student's response, discuss it, or use it to answer another student. Rosenshine[83] reports similar results—a high positive correlation between teacher-direct questions and achievement—with students in an elementary classroom and not in science. Although Riley[84] reports that teachers can be helped to acquire skills in asking more higher-level questions, the question of whether science teaching outcomes are enhanced or hindered by teacher questions remains unanswered by the research to date.

Thus, a teacher who wants to nurture positive student on-task behavior and science achievement by providing appropriate response opportunities should—

1. Keep a clear focus in the discussions in the science classroom by being businesslike and direct.

2. Use student ideas, especially by applying them in a new context or by comparing the content of two students' contributions.

3. Use fewer rebukes.

4. Use more low-level questions and fewer high-level questions.

5. Use a longer wait-time.

6. Provide frequent feedback to students.

SUMMARY

In total, what do we know about the relationship among teaching behaviors, student study behaviors, and student outcomes that is valuable to science teachers? We certainly know less than we would like to. Yet, the results do provide a guide for teachers. In many cases, the research results confirm what some would call common sense—e.g., if the tasks are made clear, the students will learn more, and if the teacher is businesslike and task-oriented, the students will be on-task as well. But common sense or common beliefs are not always supported by research findings. Many advocates of informal schooling take little comfort in the studies that show structured classrooms to be the ones in which students achieve the most. Of course, these advocates may question the appropriateness of the measures, but that is another story too long to consider here.

Our knowledge of teaching, studying, and learning is much greater today than it was 20 years ago, and it promises to continue to increase in the future. With the aid of a conceptual model such as the one described in this chapter and continued study of the problems pertinent to teaching and learning, a review such as this written in 1990 will be able to sharpen the statements about teaching behaviors and their influence on students.

CHAPTER 4

Learning Science in Informal Settings Outside the Classroom

John J. Koran, Jr.
Lynn Dirking Shafer

THE SETTING

A recent Louis Harris poll shows that museums are top entertainment spots: The United States has some 6,000 museums which annually attract more than 300 million visitors. Major league sports garner only 70 million admissions annually.[1] Another report notes that museums draw about 40 million visitors a year for a series of "short stop" science experiences.[2] And members of the Association of Science and Technology Centers reported over 8 million school class visits in 1977 alone, while half of their annual 35 million public visits are made by people under 18.[3] Each year 160 million people visit science centers, zoos, aquariums, and a variety of other science museums and nature centers. Since this figure constitutes half of all the museum visits in the United States, "science" is clearly interesting. People seek out settings where it can be experienced. Visitors are motivated to be in these settings, and a tremendous amount of learning must be taking place in such settings. These out-of-school learning opportunities represent an untapped instructional resource!

Informal science learning occurs for a diverse population of learners in a variety of settings that are outside the classroom. Science is learned both in the family setting as students watch television, read newspapers and magazines, and visit museums, zoos, and science and technology centers, and outside the home as students join specialized clubs such as the Boy Scouts, Girl Scouts, flying clubs, bird watchers clubs, and similar organizations. The potential magnitude of science learning in these settings far exceeds that which occurs in the formal classroom setting!

There is a wide variety of characteristics that distinguish an informal learning experience from a formal one. Learners in informal settings experience no classroom structures, no time constraints, no coercive forces, and no grades. They are in an exploratory context, able to move around at their own pace and attend to those settings that attract them. Usually the differences among the learners and the differences in the amounts or kinds of things to be learned provide for "something for everyone." As in the case of museums, objects are often arranged for presentation with intents and constraints that have little, if anything, to do with learning efficiency. Few, if any, learners are ever "formally" accountable on some type of test for what they have seen or done. They are voluntary, self-motivated learners, masters of their own pace and responsible only to themselves. They frequently seem to learn despite the mode of presentation.

By way of contrast, students are required by law to participate in formal K–12 learning settings. Science content in these settings is well defined, sequenced according to some theories of learning or development, and acquired as a result of teacher presentation and student action. Within this setting each actor has a relatively prescribed role. The teacher selects the material, organizes it, presents it, presumably motivates the students to learn it, and then evaluates whether or not they have acquired the content. The student attends classes for prescribed periods of time, pays attention to what is being presented, responds when stimulated, and demonstrates achievement at the end of prescribed segments of instruction. Most teachers work hard to make the experience interesting, motivating, and profitable, and most students submit either actively or passively.

College instruction differs little from the above model, with the exception that college students, for the most part, attend college voluntarily and presumably are highly motivated to master much of what they are studying. Otherwise, both the teacher and student roles and the setting in which they are being acted out resemble the traditional teaching–learning model.

Adults who do not go on to college learn in a variety of settings and from a variety of sources. In business and the military, adults receive a combination of formal instruction and on-the-job training. In agriculture, small businesses, and industry, people learn the content and processes by a similar combination of formal instruction and on-the-job training— sometimes more of the latter than the former. In all these cases the motivation to learn or explore rests primarily with the learner.

Table 1 summarizes the differences between formal and informal learning.

THE LEARNER

As Table 1 illustrates, there is a greater diversity among learners in informal learning settings than in formal classroom settings. People of all ages, races, nationalities, and social classes have access to museums,

TABLE 1
A COMPARISON OF FORMAL AND INFORMAL LEARNING

Formal learning	*Informal learning*
1. Learning takes place in classrooms.	1. Learning takes place in museums, zoos, science and technology centers, homes, clubs, etc.
2. Learning conditions and content are prescribed.	2. Learning is through free choice.
3. Motivation is extrinsic (grades).	3. Motivation is internal.
4. The content is prescribed.	4. The content is variable and changing.
5. The content is organized and sequenced.	5. The content frequently is not organized or sequenced.
6. Attendance is mandatory.	6. Attendance is voluntary.
7. Time is standardized.	7. Each learner decides on how much time is spent.
8. All students experience all the content.	8. Many displays and objects provide something for everyone.
9. Learners are of similar ages.	9. Learners are of all ages.
10. Learners have similar backgrounds.	10. There is more diversity in learners' backgrounds.

science centers, zoos, and other scientifically rich settings; they can experience "science" individually, as a family group, as a group in a social context, or as part of a more formalized visiting group such as a school class. Some come with well-developed knowledge and skill in science, while others have little or no formal science background. As Laetch and others [4] point out, the critical factor that all of these learners have in common is "free choice." From the wide variety of displays the visitor chooses those that have personal significance for her or him.

What are some differences among people that influence what they might choose to look at or learn? In one museum study, Shettel [5] finds evidence that people who enter an exhibit situation with more science knowledge tend to learn more than those with less prerequisite knowledge. Tobias [6] has found that the greater the effort put forth in designing material to provide both information and feedback, the more likely novices or newcomers are to learn the subject. These two studies argue for providing visitors who are also there as students with necessary information both prior to and during the visit so that they will be able to relate in a meaningful way to the experience.

Another line of inquiry dealing with learning in multisensory situations such as media presentations provides the following summary:

1. General ability, measured by a variety of aptitude tests, is a good predictor of learning in a variety of settings where meaningful content is presented. Among other things this would suggest that the more able students should be permitted greater freedom to explore and learn using their own strategies, while the average and less able students should receive more guidance before, during, and after informal experience.

2. Exhibits or other informal experiences that are designed to direct attention to key objects and their most important features primarily help low-ability learners to grasp the basic ideas. High-ability students tend to learn more when they have the freedom to actively imagine, interpret, and explore possibilities.

3. Presentations with simple diagrams, figures, and symbols can be used to reduce the burden of word processing and supplement abstract interpretation to benefit students whose general ability is low or who learn better from visual information. If a teacher is aware of situations in the informal setting that require reading and abstract interpretation, prior discussion may be helpful.

4. Media training can help improve learning from a particular medium. It's important here to realize that we can help students learn in informal settings if we teach them how to attend to different methods of presentation.[7]

This same line of inquiry has resulted in the finding that learning tasks that require spatial analysis and covert manipulation of images are good for students with high spatial ability, while students with low spatial ability seem to benefit more from simple figures and diagrams that clarify meaning.[8] Again, the teacher's job might be to help students who have difficulty with complex spatial presentations to reduce them to simpler, more understandable components.

The aforementioned research findings in museum and media research provide some insight into the dynamics of learning about science in informal settings. Because learners differ in terms of a multitude of characteristics and backgrounds, various types of instructional displays and experiences will be more or less facilitative for those with different patterns of attributes. The advantage offered by many museums lies in the variety of ways in which ideas are presented. Many exhibits are set up to show relationships that become "real" for the visitor as s/he moves past or interacts with displays. Thus, in contrast to school settings, there is a possibility in the informal context of experiencing a kind of excitement that comes from seeing an idea or a thing in something like its natural relationships.

PROCESS AND RESEARCH

As people move through a museum or science and technology center or, for that matter, any informal setting, a critical factor in their learning is whether or not the exhibits capture their attention.[9] To this end, Screven[10] has explored various attention-directing methods in the museum setting that would appear to have potential in zoos, science and technology centers, and aquariums. His studies have been done with a wide variety of adult museum visitors and indicate that the teaching effectiveness of exhibits designed for a wide age range depends in large part on the careful specification of the learning outcomes desired from the exhibit. This specification results in instructional objectives or objectivelike events such as questions and commands that have the effect of focusing learner attention on what is to be learned, thus weeding out extraneous stimuli. Specifying objectives on information sheets focuses visitors' attention on key elements of a display and frequently helps them make critical associations. These studies also indicate that testlike events such as pretests or prequestions that are derived from the objectives have similar positive effects in orienting learners. Again, the pretest probably sensitizes the learner to critical aspects of the display. Other researchers who have used "programmed cards" in the same way have found that they produce more learning than low information cards.[11] How these strategies influence motivation is not clear, but they are useful for school visits because average and low-ability students require attention-directing and -focusing strategies.

Teachers can alter instructional materials in informal settings by providing information panels or written outlines before or after particularly complex experiences. As Novak[12] points out, a properly designed set of materials given to students preceding or following a visit to a museum, zoo, or science and technology center can provide them with major ideas that they can later incorporate into the concepts learned, thus facilitating learning. A number of researchers who have used questions, directions, and behavioral objectives before a learning experience refer to this phenomenon as "forward-shaping."[13] Here, information received prior to a learning experience focuses learner attention on specific relevant concepts or facts to be learned during the experience. Rickards[14] speaks of an alternative to this which he calls "backward review": after a learning experience, the students are provided with material such as questions or directions that prompt them to search back through their memories and to recall a wide range of general and specific information. In the forward-shaping case, the questions and objectives appear to produce a convergence of student attention; in the backward review situation, the learner is forced to do divergent thinking and, frequently, she or he recalls both relevant and irrelevant content and details.

In a recent study incorporating elements of the advance organizer concept, forward-shaping, and backward review, 30 high-ability seventh and eighth graders were exposed to a walk-through museum exhibit of the Florida Cave. A third of the students read an information panel describing the organisms in the cave, the zones of the cave, and the relationships in this type of habitat, and then they went into the cave. A second group walked through the cave and examined the panel when they came out. A third group was used as a control group and did not receive the information panels. Both groups that read the information panel learned significantly more than the control group. Those who read the panel before entering the cave learned more than those who read it afterward 90 out of 100 times.[15] This is a single study, but it illustrates that when teachers provide information prior to an informal learning situation, they can direct student attention and improve learning.

The most common type of study done in settings such as museums, zoos, and science and technology centers is the visitor survey. Major findings from these studies include the following: (1) Life-size dioramas and first-floor exhibits seem to be preferred by visitors at the Milwaukee Public Museum.[16] (2) Museum maps and signs have been effective in reducing disorientation and in directing traffic past a given sequence of exhibits.[17] (3) Media presentations are preferred over standard case exhibits by visitors to the national parks in Washington and Oregon, and comprehensive stories tying concepts together have been found more effective than unrelated facts.[18] Interestingly, two problems with learning in museum and science center settings have been the fatigue resulting from viewing too many objects and events in too short a time period[19] and the inclination to view exhibits for too short a period (30 seconds) and in no special sequence.[20]

Experimental and quasi-experimental museum studies conducted by Wittlin[21] confirm the findings of the aforementioned surveys. For instance, she has found that when exhibits are designed around a theme that relates the objects to one another and are given more explanatory labels, they produce more learning than when exhibits only present the objects with their names. She calls the more effective exhibits *interpretive* or *structured*. Interestingly, Wittlin's description of structure resembles the learning conditions created by behavioral objectives and advance organizers, and, taken as a package, provides a powerful rationale for teachers to create whenever possible conditions that provide direction for the average and lower-ability learners in science.

There has been little research done on participatory or manipulative exhibits and almost no research of a systematic nature with science-related content. However, Oppenheimer[22] presents convincing arguments to the effect that unless an exhibit can be changed or manipulated in some

57

fashion by a learner, it is unlikely that learning will occur. He has opined that even the so-called "discovery" methods of teaching are too directed because the learners discover only what the teacher has in mind for them to discover. These contentions are supported by the findings of several studies that indicate that people who handle specimens not only spend more time observing and asking questions about them but also learn more.[23] Participatory exhibits such as visitor-operated demonstration machines and open-ended laboratory booths have also been found to be effective in producing learning.[24]

Other studies at the American Museum of Natural History Discovery Room deal with the length and readability of exhibit labels. Here, Arth and Claremon[25] have found that single, short labels are more effective than lengthy labels and that some information about the objects is necessary initially in order to get people involved with objects. This latter observation is consistent with findings reported earlier in this chapter in that many objects in a hands-on environment are so unusual or uncommon that informal learners do not know how to *start* to relate to them. Providing a small amount of information or a question functions to prime the pump, increasing the probability that interaction will take place.

Since learning in informal settings is essentially a social phenomenon, another line of research—social learning theory—has some important implications here. Social learning theory has a long history, dating back to Tard[26] and his introduction of the idea of imitation. Later, Bandura and Walters[27] described learning through the observation of others performing in a wide variety of social situations. Children learn their male and female roles by observing their parents in these roles. They observe neighbors, relatives, teachers, and others both to refine their own role patterns and to acquire new behaviors. In the same way, apprentices learn from masters, athletes learn from other athletes and from observing their own performances, and those who are trained on the job do considerable learning through observing and modeling.

Applications of this theory have occurred in teacher training[28] and recently in learning in science museums.[29] In the latter studies, live models have been used to induce and modify the exploratory behavior of visitors while they observe a hands-on geologic change exhibit and walk through a mesic hammock exhibit. Preliminary observations of visitors in these two areas of the museum indicate the following: (1) neither children nor adults touched the various objects exhibited in the geologic change area, even though there was a "hands-on" sign present; (2) adults admonished children to keep their hands off the hands-on exhibits; (3) some adults looked closely at the "hands-on" sign and mused aloud, "I wonder what that means?"; (4) most visitors of all ages moved right past this exhibit without stopping. Similarly, in the walk-through mesic hammock exhibit, the experimenters have observed the following: (1) few people stopped at the headphone stations to listen to the descriptions of the biologic interactions in the mesic hammock; (2) more males than females stopped and looked around; and (3) visitors did not realize that the wall exhibits were related to the main exhibit and were meant to provide background information. In order to remedy these situations, the investigators trained some students to walk through these exhibits and model all of the things they wanted to see visitors do. These students stopped at headphones and listened to each of them. They talked to each other, pointing out that the commentary was different on each headphone and that the wall exhibits related to the walk-through exhibits. They stopped by the hands-on geologic change exhibit and touched and discussed the objects in it. The results were dramatic. After the live models performed in each area, participation with headphones and object manipulation increased dramatically. Visitors stayed in the areas longer and observers noted a greater diversity of visitor activities.

What is significant for science teachers is that frequently students in informal settings do not have the confidence or experience necessary to guide them in their behavior. Consequently, they are guided by previously conditioned learning behaviors that come primarily from formal learning settings—don't touch, don't stay too long, don't make noise, move along, and, basically, don't be different. A person serving as a model introduces new behavior patterns and signals that other behaviors are acceptable. Teachers can act as models; so can the more confident or inquisitive students. This approach has potential not only in museums and similar settings but also in one-on-one situations in which people are exposed to scout leaders, scientists in the laboratory and the field, and others who can model scientific skills, attitudes, and behaviors. The teacher can be instrumental in finding or creating models (through training), in making models accessible to students, and in acting as a positive model whenever possible.

IMPLICATIONS FOR PRACTICE

Previously it was mentioned that the very nature of the informal learning setting mitigates against many kinds of intervention designed to influence learning. For instance, if we make all learners perform in the same way and regulate time, space, and other constraints, we transform an informal experience into a relatively formal one. However, much of what has been reviewed here suggests that we can work with learners prior to, during, and after informal learning experiences both to prepare them and to assure that they learn the most from the experience. Similarly, we can alter the setting in a variety of subtle ways in an attempt to increase the teaching potential of the setting without damaging its informal character.

Helping the Learner

Shettel [30] and Tobias [31] have conducted two of many studies that suggest that learners must be *prepared for a learning experience.* Perhaps the single most important factor that will determine what a learner takes away from an informal science experience is what she/he comes with. In order to prepare learners for visits to museums, zoos, science and technology centers, and the like, teachers must scout the territory. While there, they must look, listen, and participate. They must watch other people behave in hands-on settings, media settings, or settings with teaching-type machines. They must ask themselves, "What does someone have to know or be able to do in order for this experience to be meaningful?" The answer should contribute to the identification of necessary prerequisite knowledge or media experiences and the specification of behavioral objectives, advance organizers, or relevant questions and cues. This planning might even include the training of students to go into the informal environment and become peer models in a variety of settings.

Cronbach and Snow [32] and Koran and Koran [33] point to a diversity in learners that goes beyond prerequisite knowledge. Their reviews suggest the need for prior instruction in a number of areas that might prepare the learner for learning in a variety of informal settings. If an informal experience such as a science and technology center visit involves being confronted with a wide variety of objects to be manipulated, the students must have preliminary experiences manipulating objects. Because free-choice situations are different from normal school learning, students need both information about how to behave in these settings and assurances that their behavior is appropriate and acceptable. Similarly, experience with learning from media such as single-concept films, regular films, slides, slides and tapes, and a wide variety of manipulable objects could improve the students' ability to learn from these. Teachers need to discuss how to pick out main points and how to relate illustrations to written materials, and to provide a model for how to address each of these different stimuli.

Finally, average and lower-ability students may require special preparation to assist them in focusing on relevant characteristics and in differentiating between the critical attributes of an object or event and those that do not account for an observed outcome. As previously mentioned, simple diagrams and figures could be used to reduce the burden of word processing in settings that appear to be too heavily verbal in nature. All of these adjustments assume that the teacher has been an astute observer, has done her or his own homework (by way of scouting the new setting), and has successfully identified what the students will be confronted with and how to help them.

Increasing the Teaching Potential of the Setting

Many alterations to the informal setting are also alterations that have been tested in the formal setting where learning researchers have found them to positively influence learning. For instance, both Screven [34] and De Woard [35] suggest that museum exhibits with attention-directing devices such as behavioral objectives, questions about artifacts, audiotape questions, or "programmed cards" tend to increase learning by focusing attention on that which is to be learned. In a free-choice situation, learners can decide whether or not they even wish to attend to these types of attention-directing devices! However, a teacher can help by explaining to students that behavioral objectives, questions, and similar devices are designed to focus their attention on important aspects of what they are experiencing. Considerable research has shown that learners frequently do not know what they are supposed to do with objectives or questions, and pass over them without profiting. Teachers can help by using questions in such a way as to provide forward-shaping and backward-review experiences before a visit to a science and technology center, during it, and after it is over. [36]

Much of the research cited indicates that visitors in informal settings tend to move rapidly, cover a lot of ground, and frequently experience mental overload and fatigue. Certainly teachers, if they accompany a group, can modulate the pace, concentrate attention through the methods suggested earlier, and, hopefully through these two efforts, reduce mental overload that contributes to fatigue. They can assure students that there is no requirement to move at a rapid pace. Multiple visits are better than one visit, if that is possible. Teachers need to discuss the nature and organization of museums, zoos, and science and technology centers, and then, after giving them an overview, encourage students to visit on a number of occasions so that they can concentrate their efforts on different exhibits during each visit. Museums and zoos provide a wonderful setting for families to share science.

The research on hands-on manipulatory exhibits suggests a number of practical things teachers can do to help their students learn in these settings. They can provide students with preliminary information about the objects they will see. If students are encouraged to handle objects, not only will their progress through the museum be slowed but also they will receive considerable sensory data. Again, students need experience *before* a museum or science and technology center visit in handling objects, characterizing them, and testing hypotheses about them. *Remember, many of the suggestions here have to do with making students both knowledgeable about, and comfortable in, free-choice environments.*

Finally, the modeling research discussed has many applications in informal settings. Surely, the teacher, other students, and parents can provide models for exploratory behaviors in informal institutional settings and one-on-one situations. Students who are less inhibited, more exploratory, more aggressive, or more interested should be designated as models by their teachers in order to encourage peer emulation. Similarly, outside of these settings and the school, the teacher is in an ideal position to identify local experts in a wide variety of fields for students to meet, to work with them on an informal basis, to ask them to appear at school, and to cultivate them as individual models in school or community clubs. Remember, in order to use modeling in this way the model must already exhibit, or be taught, the desired behavior. When the model exhibits these behaviors, attention should be directed to them. Later when students emulate these behaviors, they should be praised.

CONCLUSION

Informal learning in science is a potentially valuable adjunct to school instruction. The very nature of the informal setting and the tremendous resources available here for instruction need to be mined. Because informal learning involves institutions and people other than those found in formal settings, it is an area that can be only partially managed by the science teacher. However, if we become too involved, we can transform an informal experience into a formal school experience with at least some of the negative connotations. With thought and care, this part of the learning environment could become significant for the science education of youth. During the decade of the 80's, more people than ever before will visit museums and zoos, and will participate in informal science. Hopefully we can transport this curiosity and spirit of discovery even farther—but we have to understand the factors that bring out these qualities. That is the challenge informal education presents to us in the decade of the 80's.

CHAPTER 5

The Effects of Activity-Based Science in Elementary Schools

Ted Bredderman

One of the central issues in elementary school science in recent years has been the degree to which *activities* should be included in the teaching of science. In this chapter the major claims made for the effects of activity-based science on elementary-age learners will be described. Then, for each claim, the evidence that has been accumulated over the past 10 or more years will be presented. It is hoped that this summation of research will provide guidance for teachers on the value of a hands-on approach to science. While no attempt will be made to tackle the questions of cost in terms of time, money, and effort, each teacher or school system has to consider these factors, once the likely effects on students of an activity science program are known. In these days of limited funds, the extra expense and time for preparation of a hands-on science program have to be justified.

ACTIVITY-BASED VERSUS NONACTIVITY-BASED SCIENCE CLASSROOMS

A number of researchers have investigated how classrooms and the behaviors of students and teachers change when an activity-based science program is used. Contrasted with the prevailing passive science programs, what do the more active methods achieve? In the hands-on approach, teacher-centered discussions and demonstrations receive less emphasis, and student-centered activities and discussions receive more emphasis. Discussions led by teachers focus on interpretation of the experiences that students have during their activities and experiments. The relative proportions of teacher and student talk are meant to shift in favor of more student discussion originating from and leading to experiments or investigations. Two widely disseminated hands-on programs—SAPA (Science, A Process Approach) and SCIS (Science Curriculum Improvement Study)—were the subject of investigations by six researchers who examined their effect on language and learning. After studying 200 activity programs and traditional classrooms, they report that the time devoted to teacher-centered talk decreases from about 80 percent in traditional classrooms to about 71 percent in activity classrooms. The reduction of talk in activity-based science classrooms occurs for both teachers and students. As expected, much of the time not being spent on talking is devoted to activity. Activities consume, on the average, about 10 percent of the time in traditional classrooms and 19 percent of the time in activity classrooms. The question these data provoke concerns the activities—if the students are talking less, are they nevertheless learning more? Do they learn both content and process to the same or greater extent in the activity programs as compared with more passive programs?

Process Versus Content

One problem with many of the studies that compare activity-based with nonactivity-based science teaching has been that the two types of instructional programs do not have the same objectives: the activity-based program puts a greater emphasis on *process* outcomes, and the traditional program stresses *content* outcomes. Often the researcher has then chosen to test the two groups of children on outcomes that favor either the activity-based program or the traditional program. Only a few researchers have tested both content and process outcomes. In Table 1 are shown four possibilities for which some research evidence is available and the number of studies of each type.

Combining the Results of Studies

While the content of the programs is varied, it is possible to combine the results of studies with common characteristics to get general impressions of how well students do in classrooms where activities are used as compared with students in classrooms where activities are not used. Although the techniques for combining results are somewhat complicated, the conclusions can be expressed in a way that is easily understood. Typically, experimenters select two groups of classrooms that they believe

TABLE 1

INSTRUCTION AND TESTING IN RESEARCH STUDIES OF
ACTIVITY-BASED AND NONACTIVITY-BASED SCIENCE PROGRAMS

Emphasis of instruction			
Experimental group— Activity approach	Control group— Nonactivity approach	Outcomes tested	Number of studies
Process	Content	Process	27
Process	Process	Process	1
Process	Content	Content	15
Content	Content	Content	10

are similar in all respects except that teachers in one set of classrooms, called the *experimental* group, use activity-based science lessons and teachers in the other set of classrooms, called the *control* group, do not. Students are often tested before the experiment begins to make sure that the two groups are equal at the start. The students are then tested at the end of the experiment, perhaps a year or more later, on science processes or science content (or some other outcome), and the test results of the two groups are then compared. It is possible to combine these results from several similar studies to obtain an overall pattern of how students in the two contrasting instructional treatments compare. The question asked is: Does the average student being taught using activities remain average, or does he or she become more like one of the better students (or more like one of the poorer students) in the nonactivity group? The specific placement of the average experimental student among the control group of students for several similar studies can then be averaged. This can be done even though the tests used by the different experimenters to assess a particular outcome may not be exactly the same.

CLAIMS FOR ACTIVITY-BASED SCIENCE AND THE EVIDENCE

Teachers must determine whether switching to an activity-based science program will help or hinder student progress in specific outcome areas. The areas on which considerable research has been done are science process, science content, attitude, creativity, and language development. Most teachers would agree that these are areas that are the concern of schooling. Each area will be considered separately.

Science Process

Advocates of activity-based programs claim that science activities give the student a chance to develop generalized skills in observing, measuring, and experimenting. Activities that include the interplay of knowledge, understanding, and psychomotor behavior must be experienced. They agree that students who learn about science through a textbook are not likely to demonstrate process abilities, whether or not they have read about them or seen them used by others. For most students, these process abilities are not developed adequately through everyday experience or incidentally in textbooks or other types of programs.

What Is the Evidence? In Table 2 are shown the combined results of 27 studies involving 100 science process test comparisons between students of teachers using one of the three major activity-based programs and students of teachers using other methods of teaching science. When teachers are left to teach science on their own, to teach textbook programs, or to use combination activity–textbook programs, *the apparent net result is that these students do not perform as well on process tests as do students in activity-based programs.* Shown in Table 2 are results averaged separately for studies of (1) each of the three types of control group experiences, (2) each of the three activity-based programs, and (3) each of three socioeconomic levels of students.

Among the conclusions that can be drawn from the results shown in the table are the following:

1. When compared with text- and teacher-originated programs, the activity-based programs do result in noticeably better process abilities among the students. This result is not surprising when one considers the relative emphases on process outcomes of activity- and nonactivity-based science experiences. It does, however, support the view that process abilities can be enhanced through planned instructional activities.

2. Economically disadvantaged students in activity-based programs benefit significantly more in terms of process outcomes than do average or advantaged students. Disadvantaged students apparently must rely on the classroom experiences provided by activity-based science programs. On the other hand, advantaged students may have other opportunities to learn processes outside of the classroom, and, therefore, whether or not they are in activity-based programs does not affect their process abilities as much.

3. Of the three types of activity-based instruction represented by the three programs [open (ESS), direct (SAPA), and an open/direct

TABLE 2
SCIENCE PROCESS: COMBINED RESULTS OF 27 STUDIES OF
ACTIVITY-BASED AND NONACTIVITY-BASED SCIENCE PROGRAMS

Groups	Average placement (•) of activity-based science students among control group students (class size adjusted to 25)		Number of study comparisons
	Poorer students	Better students	
NONACTIVITY CONTROL GROUPS	ooooooooooo•ooooooooooo		
ACTIVITY EXPERIMENTAL GROUPS			
All process studies combined	ooooooooooo•ooooooooooo		100
Type of student			
Disadvantaged	ooooooooooo•oooooooooooo		30
Average	ooooooooooo•ooooooooooo		41
Advantaged	ooooooooooo•ooooooooooo		15
Activity program			
ESS*	ooooooooooo•ooooooooooo		12
SCIS	ooooooooooo•ooooooooooo		50
SAPA	ooooooooooo•ooooooooooo		38

*ESS—Elementary Science Study.
SCIS—Science Curriculum Improvement Study.
SAPA—Science, A Process Approach.

combination (SCIS)], the direct approach has the greatest effect on process outcomes. This finding for SAPA is similar to results generally found when direct approaches are compared with more open approaches in subject matter areas that lend themselves to being analyzed in terms of specific objectives. Another consideration is that SAPA is focused almost exclusively on teaching science process. This singular focus certainly allows greater attention, in the time available, to the outcomes being tested.

Science Content

Advocates of activity-based science programs argue that science content is best learned in the context of concrete experiences with the materials to which the content refers. Understanding, motivation, and retention will all be greater under such circumstances. Even in programs that stress process, it has been argued, the amount of content that is learned *incidentally* and retained will be sufficient to match that of other programs. Or, conversely, teaching science content without direct experience is so ineffective that traditional approaches, even though they stress content, will have little to show for their effort when compared with activity-based programs.

What Is the Evidence? If all the studies that test science content for the three major activity-based programs are grouped together, without regard to the type of student or the type of science approach that the control groups experience, the activity-program groups show a slight advantage over the control groups. However, this is somewhat misleading.

Table 3 shows the ranking of experimental students as compared with control students on tests of science content when the results of the studies are grouped on the basis of certain conditions. From this table it can be seen that with content outcomes, as with process outcomes, *the activity programs provide the greatest benefit for disadvantaged students and may actually handicap advantaged students on science content tests.* Activity-program students show an advantage on tests of science content when compared with students in programs of the teacher's own design, but not when compared with groups using science textbooks. SCIS students fared the best on tests of science content, especially when the content tested was that taught in the SCIS program rather than the content normally tested by common standardized science tests.

A number of investigators have recently attempted to teach the same content to two or more groups of children, contrasting the approaches associated with activity-based science and with more traditional methods. Some of the activity-based methods chosen by researchers to contrast with traditional methods are described as follows: using activities with manipulatives, allowing peer interaction, using guided discovery, using puzzles, and using conflicting situations. While these studies, as presented in Table 4, are not necessarily a clear test of all of the aspects of activity-based science, in each case they provide a contrast of at least two methods, both of which are intended to teach the same science content. Five out of the ten studies listed show an advantage for the method associated with activity-based science. Five favor neither activity-based nor traditional methods. None favors traditional methods. These results suggest that activity-based approaches may be at least as effective as more traditional approaches in teaching science content.

TABLE 3

SCIENCE CONTENT: COMBINED RESULTS OF 15 STUDIES OF
ACTIVITY-BASED AND NONACTIVITY-BASED SCIENCE PROGRAMS

Groups	Average placement (•) of activity-based science students among control group students (class size adjusted to 25)		Number of study com- parisons
	Poorer students	Better students	
NONACTIVITY CONTROL GROUPS	ooooooooooooo•ooooooooooooo		
ACTIVITY EXPERIMENTAL GROUPS			
All content studies combined	ooooooooooooo•ooooooooooooo		39
Type of student			
Disadvantaged	oooooooooooo•ooooooooooooo		3
Average	ooooooooooooo•ooooooooooooo		22
Advantaged	ooooooooooooo•oooooooooooo		6
Activity program			
ESS	ooooooooooooo•ooooooooooooo		12
SCIS	oooooooooooo•ooooooooooooo		14
SAPA	ooooooooooooo•ooooooooooooo		13
Control group science approach			
Local, teacher- controlled	oooooooooooo•ooooooooooooo		8
Text program	ooooooooooooo•ooooooooooooo		17
Activity/text combination	ooooooooooooo•ooooooooooooo		14

69

TABLE 4
STUDIES IN WHICH INSTRUCTIONAL METHODS ASSOCIATED WITH ACTIVITY-BASED SCIENCE PROGRAMS ARE USED TO TEACH SCIENCE CONTENT

Investigator and grade level	Instructional methods compared	Outcomes on which students are tested	Results
Benson— 5th grade	Pupil-investigatory or discovery method vs. traditional	Content	No difference
Billings— 2nd grade	Concrete experiences only vs. concrete experiences plus verbal instruction vs. no prescribed treatment	Content—concept of interaction and evidence of interaction	Favored concrete-experience-only group over others
Blomberg— 6th grade	Laboratory approach (ESS) vs. audiovisual approach vs. reading/lecture	Content—science understanding	No difference
Byrne— 6th grade	Activity-oriented, inquiry-based (ESS) vs. text with demonstration	Content—principles of electricity and magnetism	No difference
Davis— upper elementary	Guided discovery with materials vs. expository text	Content—knowledge of information and concepts	Favored discovery
Fuller— 3rd grade	Information from puzzle-flash cards vs. information from transparencies vs. information from written booklets	Content—concepts	No difference
Kemp— 5th grade	Activity-centered/non-verbal vs. activity-centered and verbal presentation textbook vs. textbook verbal presentation only	Content	No difference
Marlins— upper elementary	"Counterintuitive events" with demonstration discussion vs. demonstration discussion only	Content—subject matter achievement	Favored "counter-intuitive events"; no difference on later retention

TABLE 4 (CONTINUED)

Investigator and grade level	Instructional methods compared	Outcomes on which students are tested	Results
Voelker— 4th, 5th, and 6th grades	Student discovery of a generalization vs. teacher-given generalization	Content— criterion for distinguishing physical and chemical change	Favored discovery at some grades; no difference at others
Vongchu-siri—4th, 5th, and 6th grades	Discovery with objects vs. presentation with graphic displays	Science rules Science concepts	Favored discovery on rules; no difference on concepts

Attitudes

Activity-based science has often been justified to teachers primarily because it is more attractive to students, and, therefore, it should improve student attitudes toward science, science classes, and school.

What Is the Evidence? Table 5 shows a summary of the results of attitude surveys for all the studies reported on the three major activity-based programs. Again, the attitude scores have been used to rank students in activity-based program groups against those in nonactivity-based control groups. The studies have also been grouped by which activity-based program is being used and by the types of students involved. The results generally suggest that attitudes are improved by activity-based science— but the effect is not dramatic. Only a modest advantage is evident from the combined results. Johnson, Ryan, and Schroeder;[1] Metz;[2] and Davis,[3] in three different studies using inquiry and guided discovery in teaching science in the upper elementary grades, have also found advantages for these activity-oriented approaches when they assess attitudes.

Creativity

Another outcome that it is believed the activity-oriented science programs will promote is creativity. Students, when permitted to handle materials and work on problems arising spontaneously from science classroom experiences, are practicing thinking in novel ways. Trying to generate explanations, hypotheses, and experiments requires divergent thinking processes. Opportunity and encouragement to be creative, it has been argued, should bring out the creative potential in students.

TABLE 5
ATTITUDES TOWARD SCIENCE AND SCHOOL: COMBINED RESULTS OF
12 STUDIES OF
ACTIVITY-BASED AND NONACTIVITY-BASED SCIENCE PROGRAMS

Groups	Average placement (•) of activity-based science students among control group students (class size adjusted to 25)		Number of study comparisons
	More negative students	More positive students	
NONACTIVITY CONTROL GROUPS	ooooooooooooo•ooooooooooooo		
ACTIVITY EXPERIMENTAL GROUPS			
All attitude studies combined	ooooooooooooo•ooooooooooooo		22
Type of student			
Disadvantaged	ooooooooooooo•ooooooooooooo		7
Average	ooooooooooooo•ooooooooooooo		13
Advantaged	Too few comparisons		
Activity program			
ESS	Too few comparisons		
SCIS	ooooooooooooo•ooooooooooooo		13
SAPA	ooooooooooooo•ooooooooooooo		7
Attitude toward school and teachers in open and traditional classes[4]	oooooooooooo o•ooooooooooooo		17

What Is the Evidence? Table 6 shows the results of tests of creativity for all contrasts of the three major activity-based programs. Studies on creativity have not been reported for advantaged students. The results in general parallel those for attitudes. The activity programs show an advantage over other approaches, and, of the three activity programs, ESS is the most effective in this area.

TABLE 6

CREATIVITY: COMBINED RESULTS OF 5 STUDIES OF
ACTIVITY-BASED AND NONACTIVITY-BASED SCIENCE PROGRAMS

Groups	Average placement (•) of activity-based science students among control group students (class size adjusted to 25)		Number of study comparisons
	Lower creativity	Higher creativity	
NONACTIVITY CONTROL GROUPS	oooooooooooo•oooooooooooo		
ACTIVITY EXPERIMENTAL GROUPS			
All creativity studies combined	oooooooooooo•oooooooooooo		36
Activity program ESS	ooooooooooo•oooooooooooo		14
SCIS	oooooooooooo•ooooooooooo		7
SAPA	oooooooooooo•oooooooooooo		15
Creativity in open and traditional classes[5]	oooooooooooo•ooooooooooo		11

Language Development and Reading

Despite the fact that none of the three major activity-based programs relies on textbooks and, in general, reading is kept to a minimum, many educators have argued that activity-oriented programs should have a positive effect on language development and reading.[6] They contend that language and reading are dependent as much on broad, meaningful experiences, interests, and general communication opportunities as on instruction directed at language development or actual practice in reading.

What Is the Evidence? With the exception of disadvantaged student groups in which the effect is strongly positive, the activity-based programs, on the average, appear to have only moderately positive effects on language development skills, as can be seen from Table 7. This modest,

TABLE 7
LANGUAGE DEVELOPMENT AND READING: COMBINED RESULTS
FOR 10 STUDIES OF ACTIVITY-BASED AND NONACTIVITY-BASED
SCIENCE PROGRAMS

Groups	Average placement (•) of activity-based science students among control group students (class size adjusted to 25)		Number of study comparisons
	Lower scores	Higher scores	
NONACTIVITY CONTROL GROUPS	ooooooooooo•ooooooooooo		
ACTIVITY EXPERIMENTAL GROUPS			
All language development and reading studies combined	ooooooooooo•ooooooooooo		49
Type of student			
Disadvantaged	ooooooooooo•oooooooooooo		12
Average	ooooooooooo•ooooooooooo		25
Advantaged	ooooooooooo•oooooooooo		8
Only studies with control groups using science textbook programs	ooooooooooo•ooooooooooo		29
Language area			
Reading	ooooooooooo•ooooooooooo		33
Listening	ooooooooooo•ooooooooooo		5
Expression	ooooooooooo•ooooooooooo		11

but significant, advantage is held even when only the studies in which the
control group is using a textbook program are considered. This last finding
presents counterevidence to the often heard argument that switching to an
activity approach removes a support for the school's reading program
because reading from subject area textbooks will be decreased. Not only
does it appear that lessening the dependence on science texts may not have
a harmful effect on reading skills but also, as shown in Table 7, it seems
that increasing the use of activities may strengthen listening and expressive

language skills. These latter benefits are found especially in studies with students at lower grade levels and with disadvantaged students.

PUTTING IT TOGETHER

The evidence presented points to the need for determining the relative importance of the many potential outcomes of schooling and for taking into account the types of students to be taught before selecting a program or approach to teaching elementary science. Certainly the several dozen investigators whose findings are summarized above have shown that teachers can expect improved performance, particularly from disadvantaged students, when using activity-based science approaches. The same may be generally true for female students, although the data on that are sparse because many of the studies fail to distinguish between the performance of males and females. It seems that with an activity-centered science program, we have a form of teaching that particularly facilitates learning for disadvantaged students — a consistent finding that has been ignored for many years. Lower ability, inner city, lower socioeconomic, and rural students benefit the most from activity-based science when compared with average ability and advantaged students. This is true for all outcome areas for which evidence is available: science process, science content, attitude, creativity, and language development. These benefits are also apparent for average students, although to a lesser extent, in all outcome areas except knowledge of particular science facts.

On content tests, students exposed to textbook approaches that stress content can be expected to outperform students in activity-based programs that stress process. The reverse is true on process tests. All things being equal, "You get what you teach for." When the content of instruction is the same for both methods of teaching, students in the activity-based classrooms generally outperform those in the nonactivity-based classrooms.

When "direct" teaching methods, as opposed to "open" teaching methods, are used with science-activity approaches, considerable benefit is shown for the specific objectives on which teaching is focused. However, the development of less definable abilities, such as creativity and improved attitudes, appears to be slightly better nurtured under more open conditions.

One thing seems abundantly clear: for those who teach disadvantaged students, a hands-on science program produces rather dramatic gains in achievement and process skills, and, as Wellman has shown in her summary, better growth in language and logic.[7]

CHAPTER 6

Attitudes and Science Education

Carl F. Berger

One of the first findings in research on attitudes is something puzzling. Students generally have a high positive attitude toward science until middle school or junior high school. Over the middle school period and through the high school years, this attitude appears to decline.[1] This is just the time when they begin to get more science in school—i.e., the number of hours required is typically 4 to 5 per week through the ninth grade. After that, for many, science is an elective, and many students do not choose to take it. What part do attitudes play in the decision to elect more science?

STUDENT ATTITUDES AND SCIENCE

Before we take a close look at students and attitudes, perhaps we should know what we're looking for. Over a 20-year span, the characteristics related to a scientific attitude in students have remained surprisingly constant: curiosity, willingness to change opinions, willingness to suspend judgment, openmindedness, objectivity, honesty, and rationality.[2] Accepting this broad definition of attitude, let's proceed to look at students' scientific attitudes.

Peterson has studied the changes in the curiosity behavior of students from childhood to adolescence. Contrary to the popular belief that students start school with very great curiosity and then slowly, but surely their curiosity decreases, she found that, "The sensory motor curiosity of these pupils did not decline . . . but maintained constant among individuals, and . . . the overall curiosity pupils expressed through sensory motor responses was relatively high." [3] Studying students under conditions in which they could exhibit natural curiosity, Peterson observed their reactions in a waiting room setting. While the presence of an adult helped young students slightly, older students exhibited more curiosity when left alone. [4] Natural curiosity does not seem to decline in students, but what about other measures of curiosity in science classes? The National Assessment of Educational Progress (NAEP) has gathered information on the curiosity that students express in science classes. [5] In response to the question "How often have science classes made you feel curious?" 46 percent of 13-year-old students responded "always or often"; 53 percent of 17-year-old students checked the "always or often" category. Interest or curiosity about science is widespread.

Students recognize that scientists have to keep an open mind, too. To the statement "Scientists must be willing to change their ideas when new information becomes available," over 75 percent of the 13-year-olds agreed, close to 90 percent of the 17-year-olds agreed, and more than 90 percent of the young adult population of the study agreed. Similar results were obtained from the students regarding a statement dealing with honesty: "One very important job of scientists is to report exactly what they observe."

While the NAEP data suggest that students find science interesting, some students also find that it makes them feel dumb. When asked "How often have science classes made you feel stupid/dumb?" 90 percent of the 9-year-old students said "never," 72 percent of the 13-year-old students said "seldom or never," and 62 percent of the 17-year-old students said "seldom or never." Notice that while the percentages are high and positive, they do decline as students go further in school. "How often have science classes made you feel successful?" To this question, 55 percent of the 9-year-old students, 40 percent of the 13-year-old students, and 28 percent of the 17-year-old students answered "always or often"!

What about the students' belief that science can be used to solve problems? According to the NAEP data, as students grow older there is a general increase in the belief that science can be very helpful in solving problems. Students generally believe that scientists can find cures for diseases. Far fewer believe that science can help prevent wars. When asked if they, the students, could help solve problems in the world today — such as pollution, energy waste, food shortage, and disease — 13-year-old students were less optimistic than 9- and 17-year-old students in thinking that there is anything they can do to help with these problems. It is not clear whether the 13-year-olds' dip in optimism is a result of particular school experiences in science classes or simply characteristic of this age group.

SCIENCE ATTITUDES OF TEACHERS

Teacher attitudes toward science are generally positive. Billeh and others[6] report that the more science knowledge or science courses they have taken in college, the more positive the attitudes of teachers are.

How elementary teachers feel about science if they are required to teach it is another matter. When comparing the subjects of mathematics, reading, science, and social studies, elementary teachers rate science the highest in the "not well qualified" category and lowest in the "very well qualified" category.[7] Only 3 percent feel "not well qualified to teach reading" compared to 16 percent who feel the same about teaching science. Could these feelings of not being qualified to teach science change if more science courses are taken? Or are there other factors that affect attitudes, particularly those of elementary science teachers, toward science teaching?

Attitudes held by elementary teachers about science may be influenced more by the teaching conditions and the program requirements than by science attitudes per se. Teachers report the following as major problems in teaching science: not enough money to buy supplies on a day-to-day basis; too little space for storage and activities; too little time available for classroom preparation; and numerous other problems associated with getting and maintaining supplies.[8] This sounds reasonable—but Shrigley[9] finds that making science materials and resources available to teachers "seems to do little to change attitudes toward science teaching" if one judges by actual behavior. Apparently if you are an elementary teacher, not having adequate resources legitimizes negative feelings toward science teaching, but having the resources does not change these feelings.

Several research studies indicate that the kind of science taught has a great influence on the teachers' attitudes toward science teaching and, of equal importance, on the students' attitudes toward science! During the 60's and 70's, emphasis shifted from teaching science by reading about a series of topics including demonstration activities toward a hands-on discovery approach.[10] Berger and Piper and Hough[11] have found in separate studies that teachers and pre-service teachers who are not actively involved in the teaching process prefer passive teaching techniques such as reading concepts and facts from science books. Chapter 5 in this volume by Bredderman shows that certain categories of students benefit markedly—in terms of both achievement and attitude—from an activity-centered approach to science.

Piper and Hough[12] have found that something similar seems to be the case for elementary teachers—"Positive attitudes and openmindedness of pre-service teachers were improved following a science methods course that engaged [them] in an active participatory role." Blatt[13] has noted similar results while investigating two contrasting methods of science instruction. When prospective teachers learn to teach science through some kind of microteaching experience, they are more likely to feel willing to try teaching science on their own. In addition, Earl and Winklejohn[14] report a major difference between team cooperative science teachers and those who teach in self-contained classrooms. Those who are actively involved in cooperative modes are more positive toward science teaching when compared with those who are not, even though no difference is shown in their attitudes toward science.

Interviews with teachers trying out new science programs suggest that they need to withhold judgment until the end of the second year. As many of them have said, the first time through you do not know what to expect. Attitudes toward the teaching of science tend to improve sometime in the second year of a new program. Berger[15] has found that in terms of attitude teachers who teach the Science Curriculum Improvement Study (SCIS), a hands-on science program, for two or more years cannot be distinguished from the designers of the curriculum. He notes not only that they exhibit positive attitudes toward science teaching but also that these positive attitudes are translated into particular kinds of science-related classroom behaviors. Lazarowitz's study of secondary school teachers finds parallel outcomes: "Secondary science teachers who use new programs in their teaching activities have more favorable attitudes toward inquiry strategies than nonusers; years of experience in the use of new programs is related to more favorable attitudes toward inquiry strategies."[16] In other words, the first year you try something new, there are so many unknowns that it can be frightening. The second year you know what to expect and your feelings change. This seems particularly to be the case in science where a new program often means new activities and unanticipated outcomes. It takes some courage to try a new way of doing things.

CHANGES IN INSTRUCTION THAT CHANGE ATTITUDES

A very simple change in the pacing of your science teaching can make profound differences in the enjoyment of teaching science and the quality of responses that you might get from the students while teaching science. Better yet, suppose that you were told that you could do all this just by waiting longer for a response from a student? The work of Rowe[17] and others leads to that conclusion. Rowe has found that discussions in science classes revolve largely around question-and-answer sessions. In analyzing more than 800 tape recordings made in rural, suburban, and city schools, she has found that the wait-time (that is, the time for a student to respond to a question before the teacher says something) is less than one second. If the student has not responded in this short time, the question is repeated or rephrased, or the teacher turns to someone else to respond. Not only is the average wait-time very short but also, Rowe finds, some teachers average as many as 10 questions per minute. Even more damaging is the fact that the vast majority of teacher comments on student responses are "Fine," "Okay," and "Good"—comments that are very short and highly evaluative. While it may seem that if teachers ask short questions to which students can quickly respond and if they then give warm, positive responses, a lively discussion will result, Rowe has found just the opposite. She reports that during problem solving and discussion about ideas that students are developing in science, a high rate of short questions with positive responses can have several bad effects. Among these are the following: students stop searching for deeper understanding; students do not share information, but hold it for just the appropriate moment when they can get that quick flash of reward from the teacher; and students refuse to give innovative and insightful ideas for fear that they will be wrong and will not receive positive verbal rewards. While it may sound easy to just slow down, to wait a long time for a student response, and then to ask the student to expand on what s/he has said, rather than to respond quickly with "Great," "Good," etc., it turns out to be a very difficult behavior change. One way to achieve it is to make a tape recording of a discussion in science. Fascinating still are comments from teachers who have experimented with slowing down their responses and making noncommittal responses. "I never realized what good ideas my students had." "They kept adding on to the answers of others." "I didn't feel I had to know all of the answers."

Rowe [18] reports the results of changing the average wait-time to just three seconds or longer: "The average length of students' responses increases; students initiate more responses; more students succeed in answering questions more of the time; more alternative explanations are offered; and students make more and better connections between evidence and inference." Further, she finds that teachers also benefit from the wait-time: "Teachers are more flexible in their responses to students; teachers' questions show more variability; and teachers' expectations regarding the performance of students rated as relatively weak improves."

It is small wonder that both the students' attitudes toward science and the teachers' attitudes toward science teaching improve. It is surprising that just increasing the wait-time when asking questions and decreasing the simple, direct positive rewards can make a great difference in student and teacher attitude.

What other behavioral changes can be made to change teacher attitudes? As Chapter 5 has shown, a hands-on science program is particularly beneficial for disadvantaged elementary school students. If teachers use such a program, the students do better. This, in turn, improves the attitude of the teacher who, in turn, presumably teaches better. For example, Johnson [19] concludes a study with the following comment: "It is clear from this study that sixth grade students who interacted with concrete materials (batteries, bulbs, wire) to answer questions developed significantly more positive attitudes than sixth grade students studying similar subject matter from a sixth grade science book." Other statements also reflect this change:

"It appears that the Science Curriculum Improvement Study (SCIS) program helps to create a classroom environment that is conducive to self-concept development. . . . Perhaps programs such as SCIS help elevate the way a child feels about himself by encouraging creativity, innovation and independent thinking." [20] "The students' perception of the learning environment does appear to correlate to student attitude towards science." [21] "Despite the fact that previous research has been mixed in regard to the effect of curriculum, it was clear that pupils in this school preferred classroom formats that stressed active involvement and experience." [22] In a study that stresses the use of experiments, "Students are not as frightened of Interdisciplinary Approaches to Chemistry, they are having more fun, they dislike chemistry less, it holds their interest more and they seem to understand more." [23] "The individualized approach to materials presented may account for the positive change in attitude of the children in the experimental biology program." [24]

In a society becoming more conscious of the impact of science and technology, we as teachers must do all we can to develop positive attitudes toward science. Not attitudes of science being or having the answers but an attitude toward science of withholding judgment, constant curiosity, openmindedness, and objectivity. A world that requires more rather than less specific solutions must expect solid science attitudes among the populace.

CHAPTER 7

The Role of Laboratory Work in Science Courses: Implications for College and High School Levels

Elizabeth H. Hegarty

For much of this century, educators have regarded laboratory work as the hallmark of science-based courses offered at the university level. This suggestion of uniqueness has probably prompted the debates and the large numbers of research studies on laboratory work listed in the compilation by Champagne and Klopfer [1] covering the period from 1916 to 1976.

During the 1950's and 1960's, however, traditional laboratory work was widely criticized as a meaningless ritual and a waste of time.[2] If, in the wake of such criticism, we ask the question, "What does research say to university and high school teachers concerning the role of laboratory work?" a ready answer might be, "Far too little and that not in an accessible form." Boud and others [3] have criticized the lack of cross-referencing and acknowledgment among authors and the lack of cross-fertilization among "education research" journals and "teaching" journals. For example, use of the audiotutorial laboratory method has remained primarily an innovation in biology teaching at the university level,[4] whereas use of the Keller plan,[5] with or without laboratory work, is chiefly restricted to physics and chemistry.

This chapter is an attempt to bring together theory, research, and strongly based descriptive studies in curriculum, teaching, and learning in order to help people make decisions related to laboratory experiences in college and secondary science programs. It is intended for teachers and course planners who are grappling with questions concerning the need for laboratory work, the form it should take, and the cost–benefit considerations which are essential, given the current state of our economy. The chapter's main contribution to thought about science education in the 80's will be to aid in the decision-making process.

A WAY OF VIEWING THE ROLE OF SCIENCE LABORATORY WORK

A model of the educational process in science (Figure 1) shows the place of laboratory work in a schema that extends earlier models[6] in several ways. It recognizes the potential for mismatch among the plan of a lesson, the intended program, and how the lesson works out in the actual program. It calls attention to those bodies of experience ("commonplaces"[7]) upon which sound judgments will depend. It illustrates the constraints that play an important role in deliberation or negotiation when institutions plan curriculums internally or implement the work of outside planning groups.

The model helps us address questions explicitly that we might otherwise ignore or answer by default. For example: (1) What things are better learned in the laboratory and what should be left to nonlaboratory procedures? (2) How do teachers choose intended learning outcomes (ILO's) for laboratory and nonlaboratory classwork? (3) To what extent do the actual learning activities (ALA's) reflect the intended learning activities (ILA's)?

Questions of "degrees of fit" focus on the *arrows* in the model. The research literature shows that many of the research questions and results fit best in the *boxes* of the model. It is easier to answer the question, "What are the goals of laboratory work (its ILO's)?" than to answer the question, "What should the instructional plan be for each type of ILO?" To date, we have been able only occasionally to probe different bodies of literature and begin to answer the *arrow* questions. The model will serve as a way of conceptualizing the knowledge accumulated about the role of laboratory work, as well as a way of conceptualizing the interactions that may be the subject of future research.

LABORATORY CLASSWORK GOALS FOR SCIENCE MAJORS, NONMAJORS, AND HIGH SCHOOL STUDENTS

We will consider six types of goals (a modification of the scheme proposed by Klopfer[8]):

1. Knowledge and comprehension
2. Manual skills
3. Processes of scientific inquiry
   a. Observing and measuring
   b. Interpreting data

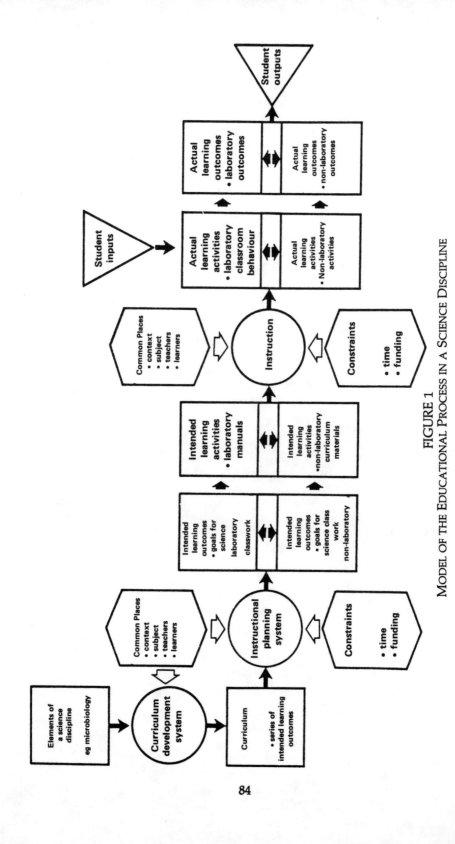

FIGURE 1

MODEL OF THE EDUCATIONAL PROCESS IN A SCIENCE DISCIPLINE

c. Identifying problems

d. Seeking ways to solve problems

4. Appreciation of the ways in which scientists work

5. Scientific attitudes

6. Application of scientific knowledge and methods

The relevant bodies of literature (on the selection of appropriate goals for university science majors and nonmajors and for high school students) provide emphasis on each of the six types of goals.

A major research study on the goals of college science majors and nonmajors and of public secondary students, Fraser's 1976 work is one of the most comprehensive of its kind.[9] It shows that the differences in majors' and nonmajors' programs are a matter more of coverage than of emphasis. Secondary schools focus on knowledge, comprehension, and an appreciation of the ways in which scientists work. For the college nonmajor, instructors emphasize knowledge, comprehension, and the application of scientific knowledge and methods. In addition to knowledge and comprehension, educators find it important for them to acquire laboratory skills and certain attitudes and processes that would help them function in science.

GOALS FOR LABORATORY VS. NONLABORATORY CLASSWORK

"What goals is laboratory work best suited to achieve?" The outcomes of comparative learning research studies are pertinent here. These studies of the effectiveness of laboratory work have been of the Method A vs. Method B variety—e.g., outcomes of instruction by lectures/discussions/demonstrations vs. instruction by individual laboratory work. If the results of a large number of studies are accumulated, the findings can provide guidance for the instructional planners who are trying to answer the question, "For which goals is laboratory work the more valuable or even the superior mode of teaching?" From reviews,[10] we can draw two major conclusions:

1. Laboratory work is valuable for teaching manipulative skills, increasing understanding of the apparatus involved, designing experiments, developing scientific attitudes and laboratory resourcefulness, fostering certain problem-solving abilities, giving practice in processes of scientific inquiry, and providing for individual differences.

2. Lecture/discussion/demonstration is superior for the presentation of complex material and more efficient for the presentation of large amounts of factual information and concepts.

Educators consider many goals other than knowledge as being important for science majors. These appear to be better taught via laboratory classwork. For nonmajors, knowledge and the application of knowledge are of prime importance. For them, it would appear that laboratory work is largely superfluous and should be discontinued forthwith. But we must pause to see what has been meant by *nonmajors*. In the training of doctors, for example, Kogut writes:

> Sticks and stones . . .! . . . It won't surprise anybody, I should think, to hear that in most of these [fifteen Canadian and U.S.] medical schools laboratory practicals in biochemistry have been simply abolished, or "withered away." [11]

It is important to note that the findings of comparative learning research are usually associated with examinations that demand recall of knowledge as the main criterion in comparing "laboratory" and "nonlaboratory" groups. Learning theorists, and especially Ausubel,[12] should prompt us to ask of this research how much of the learning is rote, without understanding, and how much is meaningful. Probably one of the most powerful uses of the laboratory concerns its effectiveness in removing or correcting students' misconceptions, most of which originate in their out-of-school experience. (See also the discussions of misconceptions in Chapters 8 and 9 by McDermott and Minstrell, respectively.)

INSTRUCTIONAL PLANS TO MATCH LABORATORY CLASSWORK GOALS

Knowledge and Comprehension

Laboratory classes are not efficient for the presentation of factual information. Reception of such knowledge via texts or lectures is superior. However, as new concepts are introduced, laboratory work or demonstrations may be necessary to give meaning to the concept as a "label." The Keller plan, or PSI (Personalized System of Instruction),[13] and AT (the Audio-Tutorial Approach)[14] are two instructional plans that seem well matched to the goal of joining knowledge and comprehension. Both were developed to address the special problems of the large numbers of nonmajors enrolled in freshman science courses at the college level. Both approaches allow for self-pacing and flexible timing—which probably accounts for consistent reports of good student acceptance. Comparative learning studies tend to indicate that AT and PSI are at least equal to the "conventional" lecture/laboratory in terms of student knowledge gained and retained.[15] For AT and PSI there are both pluses and minuses. On the plus side, students do at least as well as those in conventional programs, and there is better use of facilities and faculty time. On the minus side, the course designer may be too isolated from the teaching process; student procrastination and difficulties in controlling the testing system pose additional problems.[16]

The AT approach focuses on the integration of theoretical and practical material. It almost always involves laboratory work which students do in individual carrels. The AT approach has been used chiefly for the study of botany, zoology, microbiology, and general biology by nonmajors,[17] and also for the study of plant anatomy by more advanced university students.[18] There are signs that educators can take the audiotutorial approach beyond biology to include other sciences—e.g., chemistry[19] and physics[20]—with about the same results.

By comparison, a PSI course with a laboratory component is the exception rather than the rule, and instructors have limited its use to the physical sciences, especially chemistry[21] and physics.[22] This limitation is probably due in part to tradition and the lack of communication among teachers in different disciplines. However, it is possibly also due to the emphasis in PSI on mastery learning. In principle, this means that all laboratory materials would need to be available throughout the entire course. This is less likely to cause organizational problems in the physical sciences than in the biological sciences where animals, plants, microorganisms, and biologically active extracts have limited life spans.

In an attempt to investigate the match between student aptitudes and instructional plans, Ott and Macklin[23] have conducted aptitude–treatment interaction study involving 575 students enrolled in an AT freshman physics course at Cornell University (91 percent are engineering students and the remainder are physics majors). When they use course grade (a composite of marks for laboratory reports, quizzes, and formal examinations) as the measure of achievement, there is evidence of significant interaction between aptitudes and treatments. The authors conclude that lecture/recitation/laboratory favors higher ability students, possibly because of the greater amount of time allocated to recitation which emphasizes problem solving. The audiotutorial/laboratory favors lower ability students, possibly because of the greater degree of individualization, the availability of a wider range of instructional media, and the careful sequencing and structuring.

Manual Skills

It may seem intuitively obvious that laboratory classwork provides an excellent setting for the teaching of manipulative skills in science. However, one conclusion that we can draw from the survey and research literature is that educators seldom regard the mastery of manipulative skills as having great importance for any group of students except perhaps for college science majors. Otherwise, they tend to regard laboratory skills as process goals—i.e., means toward other ends.

Thus, the general question for this section becomes, "What intended learning activities are appropriate when the learning of manipulative skills *is* regarded as an important goal?" The briefest answer, drawn from studies of adult learners, is that the learning of technical skills, such as performing laboratory manipulations and using laboratory equipment, requires practice. Learners continue to improve their skills with practice over long periods and, once learned, they retain these skills well.[24]

For nonmajors, the rationale seems to be that it doesn't matter if students' laboratory techniques are poor; the majority of them will never use those techniques again. However, we can argue that there are *some* techniques that will be of value to, say, medical students in their later medical practice and that these can be well taught in laboratory classes.[25] We should be careful to identify such techniques and allow time in laboratory classes for their learning and practice.

For science majors, the rationale for emphasizing laboratory skills seems to be that these students should have a good repertoire of techniques and familiarity with common equipment, and that these techniques should be accurately performed. Students and educators differ in what they think is important about laboratory work. Boud[26] in physics and Lynch and Gerrans[27] in chemistry have found that while students rate technical skills, especially those of vocational importance, as having high ideal importance, staff often give them lower priority.

Runquist[28] describes the development and use of technique kits for university chemistry students. These have been well accepted by the students who apparently believe that the acquisition of such techniques is "relevant." After a student uses the kit to practice a technique, he or she completes an exercise that requires precise execution of the technique — e.g., weighing, titrating, preparing standard solutions, and using a spectrophotometer. The course is very effective in achieving its two principal objectives:

1. To reduce the faculty laboratory teaching load by at least 50 percent (from 16 to 8 hours per week)

2. To improve student laboratory techniques from an estimated initial accuracy of ±5 percent relative error to less than 1 percent in basic techniques.

Beasley[29] has investigated the effects of physical and mental practice by university chemistry students on the skills required in volumetric analysis. First, the students review the "executive subroutines" for use of the analytical balances, pipettes, and burettes that are explained and illustrated in booklets distributed to both the "physical practice" and the "mental practice" groups of students. The physical practice booklets require students to complete skill-practice exercises in the laboratory, while the mental practice booklets contain structured exercises whereby students mentally rehearse the steps of the procedure. In the experiment, all three groups (physical, mental, and physical plus mental) performed significantly better in terms of accuracy and precision than did the control group. This demonstrates the feasibility of improving the laboratory skills performance of freshman chemistry students through planned skill-practice activities. The author concludes that the most interesting implication for science laboratory instruction is the equal effectiveness of mental and physical practice. Mental practice activities impose no great burden on laboratory resources and can be easily sequenced as prelab exercises.

Processes of Scientific Inquiry

Understanding both the processes of scientific inquiry in a discipline and the development of requisite scientific thinking skills is an important goal of science instruction, especially for science majors. There seems to be little point in reiterating the supporting arguments here; they are amply covered in many sources (the cumulative index to *Science Education*, Volumes 1-60, 1916-1976, edited by Champagne and Klopfer,[30] is especially informative). The research associated with current theories of learning—e.g., those of Ausubel,[31] Gagné,[32] and Piaget[33]—tends to support the role of the laboratory in this development. However, there is constant criticism of the types of learning activities provided. The criticisms include failure to provide a meaningful context for learning as well as failure to provide opportunities for students to experience the processes of scientific inquiry. For example:

> Teachers tell students too much; they deprive them of the opportunity to learn for themselves. In the laboratory, for example, they are likely to tell them just about everything—how to assemble an apparatus, how to design an experiment, and what outcome to expect. Of course, they think they do this for a good reason—to save time and to "save the experiment." However, it seems that in spite of their enthusiasm, planning and zeal, the students frequently leave the laboratory having performed the exercise well, but with low retention of information and even lower comprehension of the significance of that information.[34]

Overall, theory and research suggest that laboratory instruction must meet three major requirements:

1. Students cannot conduct meaningful inquiries in areas in which they have no background. Course planners should design activities that provide for prior learning of the basic concepts and laboratory skills that will be required.

2. If students are to conceptualize the processes of scientific inquiry as conducted by scientists, there must be *explicit* instruction on the topic, as well as any *implicit* instruction that may be embedded in inquiry / discovery-oriented laboratory exercises.

3. If students are to experience the processes of scientific inquiry, course planners must design special learning activities. Laboratory "cookbooks" are not effective.

Hill[35] has shown that the addition of special audiovisual materials can encourage creative thinking in the laboratory (in an introductory chemistry course at the university level). Examples of tests for creative thinking in laboratory situations include those designed both to examine the problems that might arise in connection with laboratory materials and to suggest multiple acceptable methods of solving laboratory problems. There appears to be an overlap between the concepts of "creative thinking in the laboratory"[36] and "laboratory resourcefulness" used earlier.[37]

Wheatley[38] set out to show that scientific thinking skills *can* be successfully taught and evaluated in the science laboratory. The investigation was carried out in Biology 100 at The Ohio State University, a course that was taught using an audiotutorial/laboratory approach. All students taking Biology 100 experienced the full course. In addition, an experimental group undertook a series of seven special learning activities involving collection of original data, analysis of the data, generation of hypotheses based on the data, and evaluation of these hypotheses in terms of either new data presented to the students or other hypotheses that also account for the data. Three posttests spaced among the seven special activities indicate that a significant difference in performance above that of the control group is not achieved until all seven activities have been scheduled. Furthermore, students who complete less than four of the seven special activities show no improvement over students in the control group. This finding is of interest in view of the Piagetian notion that constant challenge may be necessary to promote the development of certain intellectual skills. In view of Gagne's work on the hierarchical nature of intellectual skills, it would be useful to know if Wheatley's seven special activities were arranged in the direction of a hierarchy of intellectual skills or whether all were supposedly teaching the same range of intellectual skills. The lack of improvement by students undertaking less than four of the seven exercises would be congruent with a hierarchical arrangement.

Basically the studies by Hill[39] and Wheatley[40] show that educators who provide additional, specially designed instruction will be successful in promoting their intended learning outcomes.

Investigations have been undertaken to see whether instruction can improve the science process skills of science nonmajors and of premedical and predental students.[41] A one-semester course in science process skills, based directly on the school curriculum Science, A Process Approach, has been developed and offered to small groups of students at the Southern Illinois University Medical School. As a result, there are, overall, modest improvements in the skills of measuring, quantifying, and inferring although no significant gains are made in observing, classifying, experimenting, and predicting. Using a locally made laboratory course designed to give instruction and practice in intellectual skills in chemistry for prenursing students, Ophardt[42] has reported similar modest gains.

Pavelich and Abraham[43] report the development of guided inquiry laboratories for a large general chemistry course for science majors. The laboratory program is designed to accomplish three main goals: (1) to acquaint the students with the fundamental laboratory techniques and procedures required of science majors, (2) to give the students experience with aspects of scientific inquiry, and (3) to enhance the students' abstract thinking processes. The resulting papers provide some interesting examples of the way in which "verification" exercises can be converted into "guided discovery" exercises and of the use of Explanation–Invention–Discovery cycles.[44] Q-sort data show that the students in the experimental group perceive the laboratory as having more of an inquiry orientation than do the students in the control group—e.g., students in the experimental group perceive a greater emphasis on the requirement that students use evidence to back up their conclusions and to explain why things happen. Other results are less helpful because testing for the outcomes of goal 2 (scientific inquiry) and goal 3 (formal reasoning) has been carried out simultaneously, using only Piagetian tests that show little or no improvement over the time of the course.

An elite group of 30 Yale freshman physical chemistry students participated in a small study on the ability to design an experiment investigating heat capacities, given only limited guidance and tables of molecular weight solubilities and enthalpies.[45] The most interesting finding is the identification of three types of intellectual styles among the students, as judged from their written plans for their experiments. "Empiricists," "borderline," and "dead-reckoners" emerged. The small group of "empiricists" far outperformed the other students, and the authors conclude that they behave in a manner congruent with scientific inquiry, while the "dead-reckoners" develop only "recipes." Some quotes from student papers typical of the three types show what is meant:

Empiricist:	"If temperature jump is very small ($< 2\,°C$), add more salt next time."
	"If small jump on first run, use less water on second run."
Borderline:	"Do calculations before going on to see if you are messing up blatantly."
Dead-reckoner:	"Record all measurements at steps underlined in red."

THE USE OF COMPUTERS

The wider availability of computers and the development of microcomputers mean that questions of when to use computers (and when *not* to use them) may become issues to be taken more seriously in science education in the 80's than they have been to date. The overall question we seek to answer in this section is the same as in the previous one—how to match intended learning activities to intended learning outcomes for laboratory classwork.

Much has been written over the last decade on computer simulation of science laboratory work, especially at the university level (Boud and others[46] have recently published a review of this work). Many developments have been criticized on the basis that such simulations merely repeat what is done by more traditional methods—but at higher cost (although the development of microcomputers may change the cost differential). Educators now consider it undesirable to use computer-assisted instruction as a complete substitute for any other teaching method, including laboratory work. However, it can be used very effectively in conjunction with the laboratory in certain clearly defined areas. The main benefits to be had are in the simulation of real situations that are difficult or impossible to achieve in traditional laboratories. Computers can provide practice in two very important processes of scientific inquiry: interpreting data and seeking solutions to problems.

An example of data interpretation would be the simulation of NMR spectrometry data, which is otherwise time consuming and expensive to obtain.[47] However, the advantages of using a computer over using a "dry-lab" format for conceptually similar tasks (e.g., Raman spectroscopy[48]) are not clear. The advantages seem clearer in situations in which the computer is used to store huge amounts of data of a type already familiar to students, thus saving considerably on time and cost while providing practice in interpretation—e.g., spectrophotometric determinations of pK values by university chemistry students[49] and interpretation of physiological data of numerous kinds in the MAC family of models used at McMaster University Medical School.

Computers can aid students in designing experiments and planning modes of action. Different approaches can be tried, different bodies of data generated by the computer, and different analyses obtained. Special benefits accrue in disciplines in which extensive statistical analysis is required.

A third, and apparently inappropriate, use of computers has been in teaching the operation of some laboratory instruments on the following basis: if using the instrument involves simple skills (such as turning a knob), which it is presumed that a student possesses to some degree, then the student may be able to learn to use the instrument by learning the sequence of operations required for its use. Research studies involving instrument use in a university course in chemical practice[50] and a Navy training school[51] have shown no advantage to using computers for teaching technical skills. It would appear that the potential function of a computer in teaching "executive subroutines"[52] is confounded with the need to teach physical manipulations in those studies. But even if this were not so, it is difficult to imagine great advantages for a computer under these circumstances.

INTENDED LEARNING ACTIVITIES COMPARED WITH LEARNING OUTCOMES

As the model in Figure 1 shows, analysis of the content of the written curriculum materials that describe intended learning activities can provide systematic evidence by which to judge the suitability of the materials for use in specific situations. Judgments can be based on the relationship between intended learning activities and intended learning outcomes or the relationship between intended learning activities and actual learning activities.

To date, educators have rarely analyzed the content of written materials in science education; it has been described as a promising, but neglected, technique.[53] Provided staff have a clear idea of the goals they hope to reach, content analysis permits systematic examination of materials and, if used as a monitoring device at the pretrial stage, could save time, money, and effort. Clarke[54] has reported a pretrial analysis of laboratory-centered written materials for secondary school science. When the content of a course that is already in operation is analyzed, attention can be directed toward areas in need of revision. Content analysis also provides a data base for testing and evaluation procedures.

A good example is the scheme for content analysis developed by Herron[55] (shown in Table 1), which defines levels of scientific inquiry according to the degree of student involvement both in identifying the problem to be investigated in a laboratory class and in designing the materials and methods for the investigation—i.e., the scheme depends on the dimension of guidance in scientific inquiry (discussed at length by Shulman and Tamir[56]).

The design of experimental methods is central to the whole notion of scientific inquiry and is represented as level 2 in the scheme shown in Table 1. Therefore, teachers who claim that scientific inquiry is an important goal of laboratory work can easily tell if their plans measure up. If no laboratory exercises can be found that are at level 2 or above, then it is unlikely that such a course provides much opportunity for scientific inquiry in the laboratory.

TABLE 1

SCHEME FOR DETERMINING LEVELS OF SCIENTIFIC INQUIRY IN LABORATORY MANUALS [57]

Level of inquiry	Definition of Level		
	Problem	Ways and Means	Answer
0	Given	Given	Given
1	Given	Given	Open
2	Given	Open	Open
3	Open	Open	Open

At the lowest level in the scheme are exercises that either provide practice in techniques or are confirmatory exercises with the answer already provided for students. An example would be the familiar physics exercise in which the aim is to "prove" Ohm's Law, and the formula for Ohm's Law is given, together with full instructions and instruments calibrated on the basis of Ohm's Law. At best, such recipes provide practice in useful techniques, but educational philosophers may not classify them as "real science." At the highest level, there is laboratory work during which students have freedom to determine the nature of the problem on which they will work, as well as to design the methods and select the materials they will use.

In a study using a slightly modified version of Herron's scheme, educators who analyzed the content of some 500 exercises in nine commercially available microbiology manuals for university students failed to locate any exercises at an inquiry level higher than 1,[58] suggesting that such materials would give microbiology students extremely limited experience in the processes of scientific inquiry.

A detailed scheme for the analysis of laboratory manual content has been published by Tamir and Lunetta.[59] The first section of the two-part scheme measures the degree of integration of the laboratory work with other components of the course using four headings: precede text, integrate with text, groups work on different task and pool results, postlab discussion required. The second part of the scheme is designed largely to detect tasks related to the processes of scientific inquiry. To date, use of this scheme has been restricted to secondary school science (BSCS Yellow Version, Third Edition) where it has been found that 60 percent of the laboratory exercises require practice of techniques and 40 percent require measurement and quantitative treatment of data. About 30 percent require skills of scientific inquiry including formulation of hypotheses and predictions, design of observation and measurement procedures, explanations, transformation of results, and formulation of generalizations. Other inquiry skills are judged not to be sufficiently represented. Practically no opportunities are provided for students to design their own investigations, define problems, and work to their own design.

HOW KNOWLEDGE OF STUDENTS' LEARNING PROCESSES CAN BE USED FOR INSTRUCTIONAL PLANNING

The model in Figure 1 provides a reminder that the common-places—context, subject matter, and the needs of teachers and learners—should be brought to bear at several stages of the educational process. This section will explore briefly the ways in which knowledge of how students learn can be used in instructional planning.

This section will not focus extensively on learning theories and associated research. Because reviews abound on the significance for science teachers of research based on the work of Ausubel, Gagné, and Piaget, especially the latter,[60] it would be redundant to restate them here. Instead, the following brief recommendations on the integrated use of the findings are offered.

"Discovery" and Students' Cognitive Learning

First, the dichotomy of expository vs. discovery teaching should be discarded as being artificial and unproductive. It is not an either–or situation. The views of Ausubel, Bruner, Gagné, and Piaget can be accommodated within an approach in which the teacher introduces modern scientific concepts and then arranges opportunities for students to "discover" that new observations can be interpreted using these concepts.[61] In this approach, discovery involves the extension of concepts, a notion discussed by Strike.[62] Ausubel[63] has endorsed the approach, as have various Piagetians who recommend presentation of the undifferentiated whole, exploration (to obtain concrete experience), invention (attainment of the concept), and discovery (use of the concept in new situations)—see, for example, the work of Lawson and Renner.[64] In terms of the ideas introduced in earlier sections of this chapter, this "discovery" component of science laboratory work would need to be at level 2 or 3 of the scheme shown in Table 1—i.e., it should provide a challenge for the students to investigate either a given problem or one of their own choice, using laboratory methods of their own design. In order that the investigation be *meaningful*, the students must have command of the relevant concepts and be sufficiently skilled in laboratory techniques and use of equipment so that technical errors do not entirely overwhelm the investigation. Thus, one could imagine a series of level 0 and level 1 exercises stressing skills, concepts, and the integration of laboratory work with learning from other sources (texts, lectures, etc.)—i.e., "exploration" and "invention" exercises. Students would, periodically, be ready to undertake level 2 challenges using similar concepts and level 3 challenges moving into new situations.

A practical example of the use of a procedure similar to the Exploration–Invention–Discovery cycles[65] has been reported for an introductory chemistry laboratory course at the university level. Seeking to introduce a research orientation, Venkatachelam and Rudolph have developed a course containing repeated cycles of "learning" followed by "challenge" laboratory work (see Figure 2).[66] In the "cookbook" experiment, students become familiar with the equipment as well as with the laboratory techniques and the underlying theoretical principles. In the "challenge" cycle, students investigate an open-ended question in which their previous experience is helpful, although not limiting. The so-called divergent university physics laboratory classes[67] and an inquiry format chemistry laboratory program[68] use similar approaches.

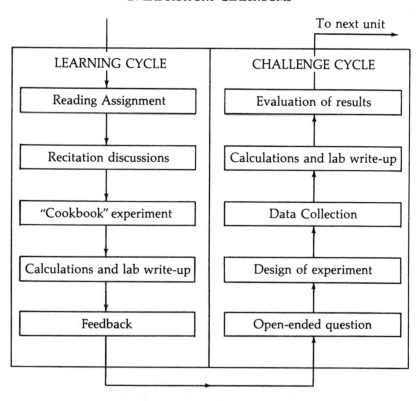

FIGURE 2
Learning Cycles and Challenge Cycles in Laboratory Classrooms

To next unit

LEARNING CYCLE	CHALLENGE CYCLE
Reading Assignment	Evaluation of results
Recitation discussions	Calculations and lab write-up
"Cookbook" experiment	Data Collection
Calculations and lab write-up	Design of experiment
Feedback	Open-ended question

Students' Cognitive Learning Styles

Research is beginning on the ways in which students' cognitive styles and personality types may interact with their learning when they are using different types of instructional materials. The Myers–Briggs Type Indicator (MBTI) appears to have relevance for research on science education; it describes profiles of students and teachers using four scales (each of which is a continuum): E (extrovert) – I (introvert), N (intuitive) – S (sensing), T (thinking) – F (feeling), and P (perceptive) – J (judging). A person could be described as preferring to use 1 of 16 cognitive styles. McCaulley's [69] research centers on understanding who chooses to enroll in science courses in colleges and what kinds of science courses and careers are appealing to students of different MBTI types. The N (intuitive) – S (sensing) scale has the highest correlation with recognized interests in science. While it appears that the majority of university science staff are of the intuitive (N) type, far more students than staff are of the sensing (S) type, and these are more likely to be science nonmajors or students in the biological rather than physical sciences. Often the S types want immediate, practical relevance in what they learn. N types appear to be more concerned with principles and understanding. The implications for teaching and learning of such possible mismatches have been discussed at the levels of both curriculum and instruction.[70]

Charlton[71] hypothesizes that sensing students enrolled in a biology course would benefit more than intuitive students from increased structure, while the latter would appreciate less structured, more inquiry-oriented activities. He has designed four different programs consisting of 10 to 15 types of learning activities for the two major MBTI types—sensing (ES and IS) and intuitive (EN and IN)—and reports the preliminary stages of a project to match students and programs. The laboratory work is organized so that sensing students undertake mostly structured exercises with clear directions given, while intuitive students are encouraged to undertake some inquiry-oriented exercises with varying degrees of specificity in the instructions as to how to get started.

Aptitude–treatment–interaction (ATI) studies would appear to be of potential value here. In the conclusions to a report of a major ATI study on school science, Eggins[72] suggests that (1) individualization is most important for students in the lower half of the intelligence range, (2) MBTI does measure personality traits that interact with teaching methods, and (3) sensing (S) students' learning is enhanced by a structured approach based on Gagné's theories while intuitive (N) students respond well to a less-structured approach based on Bruner's theories. No major S–N distinctions are reported for an approach based on Ausubel's theories (using advance organizers). Overall, it is shown that unless individualization is possible in a teaching program, no matter how you try to make instruction better for someone, you will make it worse for someone else!

Memory

Gagné and White[73] maintain that consideration of students' memory structures is important to the understanding of the effects of instruction and, at a more practical level, to the design of instruction. These authors have suggested that there may be four kinds of memory structures: propositions (concerned with knowledge and comprehension), intellectual skills (including the skills of scientific inquiry), images (visual), and episodes. The latter two seem especially relevant to laboratory work.

Images. Visual imagery results either from sensory impressions of concrete objects or events or from verbal inputs describing known objects. Thus, the student has in mind a "picture." Research studies have shown that the use of labelled drawings has a beneficial effect on intentional learning by high school biology students[74] and physics students.[75] The value of the drawings may lie in their ability to provide a summary of the text, to cue readers to certain components of the text, or to help motivate students. It would seem likely that students would obtain the greatest benefit when they are required to produce drawings themselves. In a laboratory class, students can easily be asked to make diagrams or drawings of results. The teacher's role lies in considering ways in which images might facilitate other kinds of memory structures (e.g., propositions) and result in desired outcomes; thus, students can be given guidance as to which results should be represented in the form of drawings or diagrams.

Episodes. Here memory is of an autobiographic nature. The student might recall, "First I did X, and then I produced Y." Gagné and White[76] discuss laboratory work and field trips in science, and suggest that active, colorful experiences are more likely to be stored in memory than are passive, dull experiences. Teachers should also be aware of the possible student view that much laboratory work is unmemorable, passive, and dull. Little research has been done to find out what kinds of laboratory work are memorable (White[77] refers to work in progress). Course designers may need to invent some laboratory exercises that are dramatic, emotive, or unusual. Furthermore, the "episode" needs to be effectively linked to the required technique, intellectual skill, knowledge, or concept. For example, microbiology students at the university level can be encouraged to produce food (yogurt, cheese, or bread) or alcoholic beverages (beer, wine, or mead) in the laboratory.[78] Students would then link these experiences to knowledge of the relevant metabolic processes. Ramette[79] has coined the term "exocharmic reactions" for memorable events in the chemistry laboratory; he defines the first law of charmodynamics as follows:

> Any chemical change, particularly one which is both thermodynamically and kinetically spontaneous, is inherently exocharmic and may be considered to possess an indefinite amount of latent charm.

Students' Concept Maps

There are suggestions that students may find laboratory work unmemorable because it is performed in a dull and passive manner, because they don't understand the underlying methodological principles, and/or because they fail to link it to effective conceptual structures. Current work with the "V" heuristic seems particularly apposite.[80] The "V" has at its center objects or events that are made to happen during a laboratory exercise. On the right side of the "V" are the methodological elements of knowledge-making—the records of laboratory results, their transformation into tables or graphs, and the conclusions, knowledge, or value claims that follow. On the left side of the "V" are the relevant conceptual systems, which may be represented as concept maps. These maps can be produced by the teacher or the student, but the limited research cited suggests that at the college level there may be greater effects and better acceptance of the idea if the students draw their own maps.[81]

The notion of comparing novice (student) and expert attempts at solving the same problems reveals some major differences in the strategies each uses. A study of the diagnostic abilities of medical students vs. those of practicing physicians shows the strong effect of experience when cases are typical—i.e., physicians are quicker to assimilate clues and to produce a correct diagnosis. But when cases are atypical, both groups function equally well/badly.[82]

Students' Misconceptions

Another way of learning about students' cognitive structures involves focusing more on what they either don't know or know in a way that differs from the preferred explanation in a discipline. Misconceptions can block progress. One study of high school graduates about to begin first year physics at an Australian university[83] centers on the students' understanding of gravity and involves a series of demonstrations (falling objects, bicycle wheel pulleys, etc.). Many misconceptions were revealed, suggesting that students had learned physics formulas by rote but were unable to apply them in a practical situation. Findings relevant to laboratory work include a tendency to confirm a prediction (i.e., to see what one expects to see), even when this requires amazing powers of observation—e.g., to "see" an increase in the speed of a falling eraser of $\sqrt{2}$ times over a distance of 2 meters. Because this research has proved feasible with quite large numbers of students (about 460), and because of the obvious implications for college and high school laboratory work, the protocol has been abstracted and is outlined below for those teachers who care to investigate the misconceptions of students in their own classrooms:

1. Show students the materials that will be used in the demonstration.

2. Ask questions about theory, formulas, etc.

3. Ask for predictions concerning the demonstration to follow.

4. Pose the question, "Was your prediction a guess or did you base your prediction on some knowledge? If knowledge, describe it."

5. Conduct the demonstration as appropriate—e.g., for gravity, allow a chalkboard eraser to fall beside a vertical meter scale.

6. Ask the students to describe their observations.

7. Ask further questions.

8. Ask the students to respond to the following: "If your observation is not consistent with your prediction, explain the inconsistency."

Laboratory experiences should be conducive to confronting students' misconceptions such as those reported by Gunstone and White.[84]

THE RELATIONSHIP OF INTENDED AND ACTUAL LEARNING ACTIVITIES

Missing: The Facts of Educational Life

> . . . we cannot yet describe with any confidence how teachers and pupils customarily occupy themselves. . . . In fact, we reach the overwhelming conclusion that we are spectacularly ignorant about what really transpires at the educational flashpoint, where the action is—in the classroom.[85]

According to the model in Figure 1, it is important both to observe and try to understand the behavior of students and teachers in science laboratory classrooms, and to investigate the effects of different science curriculum materials on their patterns of behavior. Although the study of classroom behavior has become established as a distinctive field of endeavor in education research,[86] the work in science education has barely begun—with relatively few studies involving science classrooms, and even fewer involving science laboratory classes, particularly at the university level.

Most of the descriptive studies concerning science laboratories at the high school level indicate that the teacher is the source of activity for as much as 80 to 90 percent of the time,[87] a proportion very similar to that in lecture classrooms.[88] However, the level of teacher talk—35 to 50 percent[89]—is much lower than the 70 percent found in lecture classrooms.[90] Figures as high as 35 to 40 percent for teachers' nonverbal, pedagogically relevant behavior[91] confirm that observation systems that categorize exclusively in terms of verbal behavior are not likely to provide an adequate description of laboratory classrooms. Examples of such nonverbal, pedagogically relevant behaviors include occasions when the teacher gives a demonstration of techniques, examines students' work, attends to routines and class management, or oversees students' laboratory work.

Dominant teacher behaviors are the development of substantive content (30 to 50 percent) and laboratory activities and organization (40 to 55 percent). Of the cognitive behaviors, there is approximately equal emphasis on low-level talk about subject matter and on talk about procedures (10 to 15 percent each), while there is less emphasis on scientific processes (7 to 8 percent). Discussion of the nature of scientific inquiry is rare, as is the reflection of scientific inquiry in behaviors such as problem identification and hypothesis formulation. Lower-level inquiry processes such as data interpretation, prediction, and formulation of conclusions are not common. Teachers seldom ask pupils to evaluate parts of the subject matter, and seldom call for new approaches to a problem or the design of an experiment to help solve a problem. (Total teacher and student behaviors in this category account for less than 1 percent of total class time.) Student cognitive behaviors recorded [92] account for about 13 percent of class time, with most of the time being taken by students both asking questions about laboratory techniques and procedures and responding to questions by providing facts and definitions.

Comparable studies at the college and university level show both similarities to and differences from the high school studies. The level of teacher talk is often higher, but it varies widely among institutions and disciplines (25 to 72 percent), while emphasis on scientific inquiry and asking "extended thought" questions is lower—generally not exceeding 4 percent.[93] The proportions of time spent on supervising students' laboratory work and on laboratory management activities are similar to those found in high school studies. The time spent by teachers on talk and activities unrelated to classwork is noticeable and consistently higher (5 to 30 percent).[94]

In comparison to their teachers, university students spend far less time (10 to 16 percent) talking and far more time (68 to 78 percent) engaging in task-related, nonverbal behavior—e.g., laboratory activities and organization. As do their teachers, university students spend considerably higher proportions of time (10 to 16 percent) than those reported in high school studies on talk and activities unrelated to classwork.[95]

The Effects of Inquiry-Oriented Curriculum Materials

Sometimes a new program does produce change for awhile. Orgren [96] has found that with the introduction of a new syllabus (New York State Regents Earth Science), there is an increased emphasis on laboratory-related activities and a decreased use of the lecture–discussion, although in all cases the lecture–"discussions" are dominated by the teachers. Studies involving BSCS (Biological Sciences Curriculum Studies) and ASEP (Australian Science Education Project) have produced some expected results—e.g., increases in student-directed activities and increases (variable) in the level of talk about scientific processes—and some unexpected results—e.g., increases in the time spent by teachers on management activities.[97] Process–product studies show a strong negative correlation between time spent on teacher management and student achievement—i.e., the class may suffer if teachers spend much time on management activities. The incidence of active behavior (e.g., questioning) is positively associated with student attitudes.[98] However, strategies that show positive effects when used with one ASEP unit do not necessarily produce similar effects with another unit.[99]

Tamir [100] has compared behavior in laboratory classrooms at high school and university levels. In addition to collecting descriptive data, he has calculated investigative indices (the sum of scores for the inquiry items in the observation system is divided by the sum of scores for the verification items). These indices reveal a dramatic difference in the inquiry behavior observed at the two levels. The average investigative index obtained for biology (BSCS) laboratory classrooms at the high school level is 1.2, compared with 0.5 at the university level—i.e., at the university level, verification dominates inquiry behavior. High scores for inquiry orientation are associated with the use of postlab discussions analyzing data and interpreting results (the time spent ranges from 7 to 29 percent). Postlab discussions are uniformly nonexistent in the university laboratory classes observed (chemistry, biology, physiology, and histology at the Hebrew University, Israel).

The failure of inquiry-oriented curriculums to materialize, especially in the laboratory, is a recurring theme in the major U.S. national report *Case Studies in Science Education.*[101] After conducting interviews, together with various types of classroom observations and anthropological approaches, the researchers offer the following reasons:

1. The amount of science content to be covered is so great that little time can be sacrificed for laboratories.

2. While students may enjoy laboratory work, especially active aspects such as dissecting, teachers encounter difficulties in ensuring that the experience either is linked to meaningful learning or can become part of an inquiry/investigation.

3. When there is overcrowding and understaffing of laboratories, "show and tell" becomes a necessary part of class control, and hands-on laboratory work is precluded.

4. Laboratory management activities (handing out, checking, collecting, etc.) occupy a disproportionately large amount of laboratory time, especially during short periods.

There have been no studies of laboratory behaviors before and after the introduction of inquiry-oriented laboratory manuals,[102] but, in a move in that direction, the effects on behavior of laboratory manuals at different levels of inquiry have been studied within a single university course.[103] The levels of inquiry of microbiology laboratory manuals are rated from low (0) to high (3) using a modification of the scheme shown in Table 1. Expected results include the increase in time spent by students talking about scientific processes (mostly with fellow students rather than with tutors). Unexpected results include the findings that this increase is not matched by an increase in the time spent by tutors and that there is an increase in the time spent by tutors on laboratory management activities (such as collecting glassware) outside the classroom (and away from students) during exercises classified as being at a high level of scientific inquiry. Inquiry-related behaviors of both students and staff are more noticeable during exercises at level 2A (structured inquiry exercises) than at 2B (projects), and at level 2B the most prominent behaviors involve organization and management. These findings represent a challenge to the practice of offering project work in university science.

As a result of studies of behavior in college and university laboratory classes, there have been urgent calls for change—in the design of laboratory exercises,[104] in the provision of postlab discussions for problem-solving investigative laboratories,[105] and in the training of laboratory teaching staff.[106] Important moves toward providing teaching training for science teaching assistants have been reported recently.[107] There is, however, no reason to suspect that problems in teaching practices are confined to teaching assistants. It may be that many university teaching staff regard themselves primarily as resource persons rather than as questioners or challengers, and that they have limited understanding of scientific inquiry or of the teaching skills likely to promote students' understanding.[108]

Workshop packages and accompanying materials on science teaching and the development of reasoning[109] are now available. Not all readers may sympathize with the use to which Piagetian theory and research have been put in these materials. Nonetheless, the emphasis on the role of the teacher as challenger and questioner is, in itself, a valuable contribution. The materials are available in different formats for teachers of biology, chemistry, and physics at university and high school levels.

CONCLUSION

What Research Tells Science Teachers About Laboratory Work

The laboratory can be a valuable setting in which students gain understanding and skill in using laboratory techniques and equipment, develop laboratory resourcefulness, learn to design experiments, prepare or select the materials needed, practice the processes of scientific inquiry in a discipline, and foster certain problem-solving abilities and scientific attitudes. In a laboratory, students can get concrete experience with the concepts and materials of a new discipline; they can have the opportunity to enlarge the conceptions they hold; their misconceptions can be confronted. Laboratory work allows them to sense and feel the dimension of science study. It can allow for individual differences in cognitive styles as well as in needs for self pacing, flexibility of timing, etc. Laboratory work can be entertaining and dramatic, but it is more likely to be of value to memory if the fun is linked to appropriate meaningful science learning.

For making instructional planning decisions, some finer connections must be made. The teaching of scientific inquiry provides an example. Activities should be designed to provide for prior learning of basic concepts and laboratory skills so that students are in a position to conduct meaningful inquiries, whether of a guided or a more open type. If it is regarded as an important learning outcome that students understand the processes of scientific inquiry (as conducted by scientists in the discipline), then there must be explicit instruction on the topic, as well as any implicit instruction that may be embedded in inquiry/discovery-oriented laboratory exercises.

Research on the current status of laboratory classwork in colleges and schools suggests that the potential is seldom being fulfilled. Where the classes could be filled with the spirit of scientific inquiry, they are dominated by procedures of low scientific status and by talk that often fails to lead to meaningful learning. Laboratory experiences are considered in isolation; students' background conceptions are not brought to bear, and their misconceptions are not confronted. Opportunities are missed for providing memorable events to aid students in their acquisition of knowledge and concepts.

It is urgent that these and other issues raised in this chapter be addressed by course designers and teachers. Ways should be sought to increase designers' repertoires for planning instructional materials at different levels of scientific inquiry (especially at the higher levels) and to help staff to develop their understanding of scientific inquiry. Teachers need to acquire an array of skills particularly suited to laboratory instruction. To achieve these and other goals, we must make imaginative use of innovations in laboratory teaching. Carefully chosen and well-sequenced combinations of laboratory and nonlaboratory work should be developed.

The Focus of Future Research

We are far from being able to answer many of those questions referred to earlier as "degrees of fit" questions (focusing on the arrows in the model presented). Research directed in this manner could help tease out answers to such questions as: "What should the instructional plan be for each type of intended learning outcome?" and "How can sequencing and balancing be achieved in order to allow fulfillment of a range of outcomes?"

Valuable current research which could help answer such questions centers on the conceptions and misconceptions that students bring to bear on their learning in science. A program of research that would enlighten educators regarding the role of laboratory work could be described as follows:

1. In any one discipline, determine the range of understandings and misconceptions held by students in key substantive areas.

2. Categorize different types of intellectual commitments, and obtain some notion of the relative frequencies of commitments in each of the categories.

3. Develop some series of learning activities (matched to categories of commitments) that are intended to create cognitive conflict

and permit students to accommodate a new conception of the subject matter. If experience has taught students misconceptions, then it is unlikely that formal instruction will overcome them. A laboratory setting could provide the link with prior experience.

4. Determine conditions under which learning activities create cognitive conflict that is taken seriously by students—e.g., find ways to produce arguments. To what extent is peer interaction required and what form should it take? To what extent are challenges by a teacher necessary, and what types of questions/probes are effective?

5. Determine the effect of the specially designed learning activities on desired learning outcomes.

Problems in Understanding Physics (Kinematics) Among Beginning College Students—With Implications for High School Courses

Lillian C. McDermott

As experienced teachers recognize, the ways in which students interpret natural phenomena differ markedly from those of the physicist. Although this problem of mismatch between scientific and "natural" concepts has been with us for a long time, it is only relatively recently that systematic attempts have been made to study the nature of students' conceptions of the physical world.

This chapter begins with a brief review of some recent studies of physics learning that have been conducted among students at the secondary and college levels. Immediately following is a description of one of these projects: an investigation of conceptual understanding in kinematics. (In Chapter 9, Minstrell describes in detail a research project on the concepts of motion in high school physics.)

CONCEPTUAL UNDERSTANDING RESEARCH

In one of the first investigations of student understanding of physical concepts, Driver and Easley[1] examined the conceptual frameworks used by seventh- and eighth-grade science students in interpreting simple mechanical phenomena. More recently, research with this same age group has been carried out by Champagne and others[2] as they have explored student preconceptions about the motion of objects in free fall.

At the high school level, Minstrell,[3] a high school physics teacher, has been conducting research on conceptual understanding among his students. He is documenting common misconceptions, monitoring conceptual development during the course of instruction, and investigating factors that appear to effect changes in student thinking.

Research on student understanding of physical concepts is also in progress at the college level. The concepts of dynamics have been the focus of studies by Viennot.[4] Under the direction of Lochhead and Clement, a group of researchers has studied the preconceptions of engineering majors in introductory mechanics.[5] Champagne, Klopfer, and Anderson[6] have examined the combined effect of conceptions of motion, mathematical skills, and reasoning skills on achievement in introductory mechanics. McCloskey, Caramazza, and Green[7] are seeking to characterize internal representations of the physical world among students with varying degrees of expertise in physics. The conceptual difficulties encountered by students in several different areas of physics have been described by Arons.[8] The Physics Education Group, under the direction of McDermott, has been engaged in a systematic investigation of the ways in which introductory physics students think about motion.[9]

These studies are representative of a field of research that is receiving increasing attention in the ongoing effort to improve science and mathematics instruction. A major goal of research in conceptual understanding is the identification and analysis of specific difficulties encountered by students in learning science and mathematics. In these investigations, the primary emphasis has been on an in-depth exploration of the thinking of individuals rather than on limited questioning of large groups. The information obtained is primarily descriptive and not readily expressed in quantitative terms. Nevertheless, the results can be generalized and shown to have wide applicability.

The emphasis on the concepts of motion that has characterized many of the physics-related studies has not been misplaced. Mechanics comprises a major part of the curriculum of virtually every introductory physics course in both high school and college. Furthermore, a sound grasp of the concepts of motion is critical in the study of almost all of physics.

A major complication encountered in teaching mechanics is that students have already acquired from daily experience ideas about velocity, acceleration, force, momentum, and work. These proto-concepts are generally somewhat vague and undifferentiated from one another, and they lack the precise operational definitions needed in physics. Often the *pre*conceptions students bring with them are *mis*conceptions that are tenaciously held and difficult to alter through conventional instructional means. Their existence poses serious obstacles to future learning of physical science concepts.

Sometimes even severe misconceptions are difficult to detect. Since many examination questions can often be answered by substituting in formulas, performance in a physics course does not always reflect the degree of conceptual understanding attained by a student. In order to examine student thinking, it is necessary to construct questions that require the correct use of a concept rather than the routine application of a formula. Furthermore, it is necessary to elicit from the student a response sufficiently detailed to reveal its conceptual basis.

CONCEPTUAL UNDERSTANDING IN KINEMATICS

As an illustration of research on conceptual understanding in physics, a study at the University of Washington will be described. This study, the first phase of a project on the concepts of motion, consists of an investigation of student understanding of the concepts of velocity and acceleration in one dimension. Descriptions of this research have appeared in the *American Journal of Physics*, [10] excerpts of which are included in this chapter. A more detailed discussion is presented by Trowbridge. [11]

Methods of Investigation

In this investigation the criterion chosen for assessing understanding of a kinematical concept is the ability to apply the concept successfully to interpreting simple motions of real objects. The primary data source is the individual demonstration interview in which students are asked specific questions about simple motions they *observe*. Similar to the clinical interview used by Piaget [12] to investigate the development of reasoning in children, this technique has proved fruitful in providing a much more detailed description of conceptual understanding than can be obtained through written means. Supplementary information is obtained from student responses to examination questions, from dialogues between students and instructors, and from class discussions. This chapter will be limited to a description of the research involving individual demonstration interviews.

Approximately 300 individual demonstration interviews were administered to students in several different courses before and after instruction in kinematics. In the individual demonstration interview, the student is confronted with a simple physical situation and is asked to respond to a specified sequence of questions. The student is asked to perform tasks that usually involve comparisons of the velocities or accelerations of two linear motions. The demonstration set-up consists of two identical steel balls rolling along a pair of aluminum U-shaped channels. A mechanism for automatically releasing the balls ensures that the motions are reproducible.

The interviews are conducted according to a specified questioning format, but at any point the interviewer may choose to probe more deeply into a student's understanding by extending the discussion. The interviews, lasting from 20 to 30 minutes each, are audiotaped and occasionally videotaped. The dialogue is transcribed and analyzed in detail.

There are several factors inherent in this type of study that must be examined for influence on the results. Among the most important are selection of the individuals interviewed, effect of learning on postcourse interviews due to participation in precourse interviews, and interscorer reliability. These factors are shown to have no significant effect on the results of the investigation.

Populations of Students

Several different student populations have participated in the study. One group consisted of academically disadvantaged students in the University's Educational Opportunity Program (EOP). These students were enrolled in a special three-quarter basic physics course intended to prepare them for mainstream science courses.[13] The course was laboratory-centered and taught in a modified inquiry-oriented manner with a great deal of interaction taking place between students and staff. The second and third groups of students came from two different sections of the non-calculus general physics sequence required for admission to many professional programs. One section was taught in an individualized, self-paced format in which students worked independently. The other section was taught in the traditional lecture format. The fourth group included students from a lecture section of the calculus physics sequence required for physics and engineering majors. Also included in the study was a fifth group consisting of participants in a program for in-service elementary school teachers.

The interview tasks were first developed during exploratory work with students in the academically disadvantaged class and later administered to other groups. The inquiry-oriented environment of the EOP class provided an ideal atmosphere for preliminary studies since pre- and post-instructional interviews could be woven into the daily schedule. All students in the class participated in interviews and seemed to respond well to the experience. Subsequent investigation established that their thinking is typical of most college students without formal training in physics. Students from other introductory courses were all interviewed on an unpaid, volunteer basis.

The Concept of Instantaneous Velocity

Two speed comparison tasks have been developed to explore student understanding of the concept of instantaneous velocity. In these tasks the students are asked to compare the simultaneous motions of two identical balls rolling on parallel U-shaped channels. In each task at least one of the balls rolls with nonuniform velocity. The important difference between the

109

tasks is that in Speed Comparison Task 1 each ball passes the other, whereas in Speed Comparison Task 2 no passing occurs.

Speed Comparison Task 1 (Passing Twice). Description of Task: In this task (Figure 1), ball A travels with uniform motion from left to right while ball B travels in the same direction, starting with an initial velocity greater than that of ball A. Ball B travels up a gentle incline, slows down, and eventually comes to rest. Ball B first passes ball A, but later ball A passes ball B. The students observe the motions of the balls, first separately and then together, several times. The accompanying graph (Figure 2), which was not used in the interviews, illustrates the motions observed.

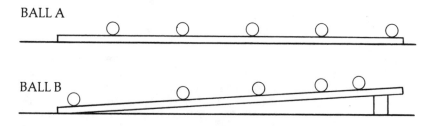

FIGURE 1
SPEED COMPARISON TASK 1 (PASSING TWICE)
(Motion is from left to right; successive positions are shown as they would appear in a strobe light photograph)

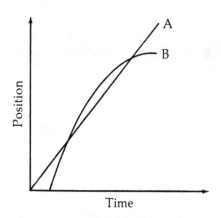

FIGURE 2
POSITION-TIME GRAPH OF MOTION DEMONSTRATED IN
SPEED COMPARISON TASK 1

110

During the course of the interview, students are asked, "Do these two balls ever have the same speed?" (The term *velocity* is also used in the interviews after it has been introduced in the course.)

Examples of Student Difficulties: We have found that a substantial number of students interpret the instants of passing as the times when the two balls have the same speed. The following response, taken from a precourse interview with a student in the calculus physics course, is typical of this student error. (The letter "I" represents the investigator and the letter "S" the student.)

I: Let's look to see whether these two balls even have the same speed. *(Balls are released.)*

S: It looks like they have the same speed twice. One is about near a quarter of a length of the incline, and then again at three quarters.

I: And how could you tell?

S: Because both balls reached the same position on each track.

I: *(Student is asked to place markers along the track beside the passing points. Demonstration is repeated three times.)*

S: Well, they both reached the mark at the same time. But before that A was traveling faster. Then after it, B is.

I: And right here, when they are side by side, what can you say about their speeds right at that instant?

S: They would be the same.

This student expresses the belief that when two objects reach the *same position*, they must have the *same speed*. He also associates being ahead with being faster.

In the following excerpt from a precourse interview with a student from the academically disadvantaged class, we again see a failure to distinguish between speed and position.

S: Somewhere around in here *(indicates region near first passing point)* they must be going about the same speed, because ball B passes ball A. So while ball B is speeding up, ball A is slowing down. There's got to be a point when they're going about the same speed.

I: You say ball B is speeding up?

S: Well, because it's coming out [released] second, and then by the time they end up, it's ahead of ball A.

This student seems to believe that when one object has caught up to another object, they must be going the same speed—even though immediately prior to the dialogue quoted above, she had described ball A as traveling with a "steady speed" and ball B as having "slowed down, so it wasn't going at a steady speed." When asked to compare the motions together, she says, "Ball B is speeding up; ball A is slowing down." She elaborates that she means "speeding up" to be catching up from behind.

For some students this apparent confusion of speed with position persists despite instruction meant to clarify the distinction. Consider the next example taken from a postcourse interview with a student in the self-paced noncalculus general physics course.

I: Now do either of those [instants of passing] represent instants when the balls have the same velocity?

S: It would only be instant. If it was more than just an instant, maybe they would go along the track side by side for a certain distance, but it's actually only about an instant that it has it.

Our interpretation of the responses quoted above is that these students lack an adequate procedure for deciding when two objects have the same instantaneous speed. Instead, they focus attention on the perceptually obvious phenomenon of passing to make the required comparison.

One might expect that once an individual has ascertained that the speed of ball A is the same during the whole trip, and that the speed of ball B continuously decreases from being initially greater to being finally less than the speed of ball A, the conclusion would be that the two balls would have to have the same speed once and only once during the demonstration. However, many students failed to draw that conclusion, and, of those who did, such logical considerations did not dissuade them from still identifying *two* instants when the speeds are the same.

Speed Comparison Task 2 (no passing). Description of Task: In the second speed comparison task (Figure 3), no passing occurs. One ball is

BALL C

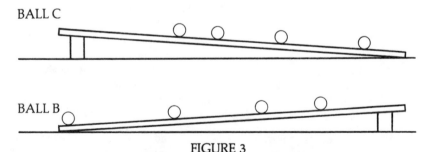

BALL B

FIGURE 3
SPEED COMPARISON TASK 2 (NO PASSING)
(Motion is from left to right; successive positions are shown as
they would appear in a strobe light photograph)

always ahead of the other ball; yet, they have the same speed at one instant during the demonstration.

In this task, once again ball B has the same motion as before. It starts with some high initial velocity, slows down, and comes to rest. Another ball, ball C, starts from rest at a point ahead of ball B. It accelerates uniformly down a gentle incline. Ball B never overtakes ball C, as is shown by the graph in Figure 4.

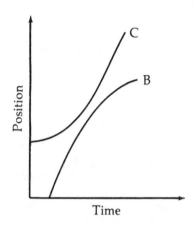

FIGURE 4
POSITION-TIME GRAPH OF MOTION DEMONSTRATED IN
SPEED COMPARISON TASK 2

Examples of Student Difficulties: It might be expected that students using a position criterion to compare velocities would claim that since the two balls in this demonstration are never side by side, they would never have the same speed. An example of this type of response is taken from a precourse interview with a student in the course for academically disadvantaged students:

I: Now, let's see whether these two balls ever have the same speed. *(Balls are released.)*

S: They don't.

I: What did you actually see happening to make you decide?

S: Well, they weren't parallel to each other so that they would be the same speed; ball C just kept on going and ball B was farther behind.

This example shows striking similarities to incorrect responses on the first speed comparison task. We see a student associating the idea of being

113

ahead with having a greater speed. The use of relative position for comparing speeds is as common on the second speed comparison task as it was on the first.

As in the case of the first speed comparison task, logical arguments do not necessarily influence student responses. Even after satisfactorily describing the speed of ball B as decreasing to zero and the speed of ball C as increasing from zero, some students observing the demonstration would still claim that the speeds were never the same since the balls never passed each other.

Analysis of Results. The two speed comparison tasks have yielded highly consistent results. Students who have difficulty with one task almost invariably have trouble with the other. Virtually every failure to make a proper comparison could be attributed to use of a position criterion to determine relative velocity.

Although this investigation was undertaken primarily as a descriptive rather than a quantitative study, it has been possible to assign numerical values to some of the data. The performance of students on the speed comparison tasks is scored on a three-point scale of 0, 1, or 2. A score of 2 means that the student has no difficulty, while a score of 1 indicates that an initial difficulty is overcome after the demonstration has been observed several times. When, after several trials, the student still has no adequate procedure for deciding when the balls have the same speed, a score of 0 is assigned. The specific scoring criteria are shown in Table 1.

Speed Comparison Task 1 (passing twice) was included in precourse interviews conducted with individuals from various introductory physics courses. The groups represented were in-service elementary school teachers, academically disadvantaged students, noncalculus general physics students in the self-paced course, and calculus physics students. Speed Comparison Task 2 (no passing) was also included in precourse interviews with the academically disadvantaged students.

Speed Comparison Task 1 (passing twice) and Speed Comparison Task 2 (no passing) were both used on postcourse interviews, with noncalculus general physics students in the lecture course included. Generally we attempted to interview the same students before and after instruction, but this was not always possible because of attrition during instruction and a conscious decision to administer only precourse interviews to one group of students and only postcourse interviews to another.

For convenience in making comparisons, the data can be reduced to a single dichotomous variable: success or failure on a task. The criterion chosen for success is a score of 2.

Prior to instruction, success on the speed comparison tasks ranged from about 40 to 70 percent. Least successful were the in-service elementary school teachers (41 percent), followed by the academically disadvantaged students (53 percent), the general physics students from the self-

114

TABLE 1
SCORING CRITERIA FOR SPEED COMPARISON TASKS

Speed Comparison Task 1		*Speed Comparison Task 2*	
Score	Criteria	Score	Criteria
0	Student demonstrates no adequate procedure for deciding when balls have the same speed. After at least three trials, student persistently identifies passing points, claims that speeds are never the same, or cannot decide.	0	Student demonstrates no adequate procedure for deciding when balls have the same speed. After at least three trials, student persistently claims that speeds are never the same or cannot decide.
1	Student initially identifies one or more passing points or cannot decide but, after successive trials, identifies a region or a time at which speeds are similar.	1	Student initially claims that speeds are never the same but, after subsequent trials, revises this judgment and identifies a region or a time at which speeds are similar.
2	Student identifies similar speeds on first or second trial without confusing speed and position.	2	Student identifies similar speeds on first or second trial without confusing speed and position.

paced section (59 percent), and the calculus physics students (68 percent). In all introductory-level populations studied, at least one-third of the students confused the concepts of speed and position during precourse interviews.

After instruction, success on the speed comparison tasks ranged from about 70 to 90 percent. The academically disadvantaged students (90 percent) and calculus physics students (92 percent) performed better than either the self-paced (73 percent) or lecture (64 percent) general physics students. In the introductory-level populations studied, overall about one-fifth of the students still confused the concepts of speed and position during postcourse interviews.

The use of the word "confused" here should not be misconstrued to mean the mistaking of one fully developed concept for another. We are saying that the confusion between speed (or velocity) and position indicates the indiscriminate use of nondifferentiated proto-concepts.

The Concept of Acceleration

The main thrust of this part of the study is to evaluate the qualitative understanding of acceleration as the ratio $\Delta v/\Delta t$.[14] From exploratory interviews it became clear that many students were aware that the concept of acceleration includes the idea of a change in velocity but did not recognize that it also incorporates in an explicit manner the idea of a corresponding time interval during which this change takes place. The students frequently referred to acceleration as a "change in velocity over time." Further probing revealed, however, that in many cases the word "over" often used by students in defining acceleration did not necessarily refer to the relationship between the numerator and the denominator of the fraction $\Delta v / \Delta t$. For these students "over" was equivalent to "during." From these interviews it became apparent that what was needed was a task that would require for its successful completion an understanding of the role of Δt as well as Δv.

Two acceleration comparison tasks satisfying this requirement were designed. One of these was administered as an individual demonstration interview; the other was presented in a written format. Only the former will be discussed here.

Acceleration Comparison Task 1. Description of Task: In this task (Figure 5), students observe the motions of two steel balls that roll down straight aluminum U-shaped channels. These channels are placed side by side and are inclined at the same angle to the horizontal. The accelerations of the balls can be varied by using channels of different widths, as shown in Figure 6. Thus, prior knowledge about the dependence of acceleration on slope yields no clues for making correct comparisons.

Both balls start from rest and reach the same final velocity at the end of the incline, just as they simultaneously enter a tunnel at the bottom. They are not released at the same point or the same time and do not travel

BALL A

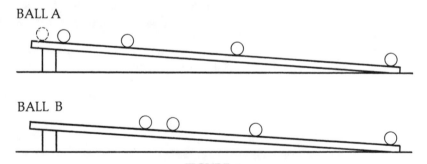

BALL B

FIGURE 5
ACCELERATION COMPARISON TASK 1
(Motion is from left to right; successive positions are shown as they would appear in a strobe light photograph; dashed circle indicates initial position of ball A; solid circles indicate corresponding positions of balls at equal time intervals)

116

FIGURE 6
CROSS-SECTIONAL VIEW OF U-SHAPED CHANNELS USED IN
ACCELERATION COMPARISON TASK 1

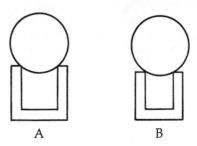

A   B

equal distances. Ball A is released first from a point several centimeters behind ball B. After rolling a few centimeters, ball A strikes the lever of a microswitch which, in turn, releases ball B. The two motions are illustrated in the accompanying graphs (Figures 7 and 8). As can be seen from the graphs, the balls have the same average velocity and the same final velocity. However, ball B, which rolls on the narrower channel, reaches that velocity in a shorter period of time than ball A and has an acceleration about 15 percent greater than that of ball A.

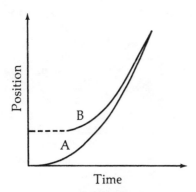

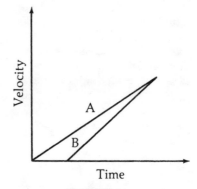

FIGURE 7
POSITION–TIME GRAPH OF
MOTION DEMONSTRATED IN
ACCELERATION COMPARISON
TASK 1
(Dashed line indicates position
of ball B from instant ball A is
released until ball B is released)

FIGURE 8
VELOCITY–TIME GRAPH OF
MOTION DEMONSTRATED IN
ACCELERATION COMPARISON
TASK 1
(Balls reach the same velocity
just as they enter a tunnel at the
bottom of the incline)

The balls are rolled separately at first, and the fact that each has an acceleration is established. During the course of the interview, the student is asked, "Do these two balls have the same or different accelerations?" Two procedures for arriving at the correct conclusion that ball B has a greater acceleration than ball A follow immediately from the definition of acceleration as $\Delta v/\Delta t$. The student must recognize either (a) that since ball A is already moving when ball B is released, the change in velocity for ball B is greater than the change in velocity for ball A between the instant ball B is released and the instant both balls enter the tunnel; or (b) that since both balls start from rest and reach the same final velocity, ball B, which is released after ball A, makes this change in a shorter period of time. Although entirely qualitative, both methods require explicit consideration of both Δv and Δt in determining the acceleration. Procedures (a) and (b) are designated as Procedures 9 and 10, respectively, in the following discussion in which the various procedures used by students on Acceleration Comparison Task 1 are summarized.

To encourage students to concentrate on the main conceptual issue rather than on subsidiary experimental details, specific guidance is provided. The interviewer explains that to make the comparison, it is unnecessary to identify the cause of the acceleration or to determine whether or not the balls, the channels, or the slopes are the same. The comparison of accelerations is to be made strictly on the basis of the motions *observed*. It is pointed out that ball A is released first and rolls for a short time before hitting the switch that releases ball B. If a student does not notice that the balls enter the tunnel at the same time or does not spontaneously compare final speeds, the interviewer asks questions that serve to direct attention to the arrival of the balls at the tunnel. Thus, the students are assisted in making the observations necessary for comparing the accelerations. It remains for them to combine this information in a manner that permits successful resolution of the task.

Examples of Student Difficulties: Acceleration Comparison Task 1 was administered both before and after instruction in more than 200 individual demonstration interviews with introductory physics students. We found that all the responses could be organized into 10 categories, each characterized by the procedure used to compare accelerations. Two of the procedures are directly based on the definition of acceleration and have already been described. The other eight, each identified by a number, are discussed below. All but the first are illustrated by excerpts from interviews. These excerpts, which have been edited to eliminate irrelevant dialogue, are typical of the responses of the students who used the particular procedure. When pre- and postcourse excerpts appear together, they are from interviews with the same student. *This juxtaposition is intended to illustrate the persistent nature of some of the mistaken preconceptions with which students often begin physics courses.*

Procedure 1—One preconception that is often valid—but is not in this case—can severely limit a student's ability to deal with Acceleration Comparison Task 1. The belief that the acceleration of the two balls must be the same because they are on the same incline was initially expressed by a large percentage of the students, especially those who had studied physics. Usually we could dissuade students from using a procedure based on slope by questioning them about various parts of the motion or by simply pointing out that the channels were not identical and suggesting that dynamical considerations be ignored. Since this procedure was almost always abandoned early in the interview, it is not illustrated by an excerpt.

Procedure 2—

I: Let's see whether we can decide whether they have the same acceleration or different accelerations. *(Balls are released.)*

S: Ball A was a little bit faster. *(It was a fraction of an inch ahead of ball B at the end.)*

I: *(After repeated demonstrations)* Ball A was released first. Then, ball B was released, and down there they had the same speed. Does that tell you anything about their accelerations?

S: I think they have the same acceleration. They ended up there at the same point.

The dialogue above illustrates the belief that when the two balls reach the same position, they have the same acceleration. The excerpt is taken from a postcourse interview with a student from the academically disadvantaged class. In claiming that when the two balls are at the same point they have the same acceleration, this student is using a position criterion to compare accelerations. This procedure is reminiscent of a similar one used by some students to compare velocities. Sometimes during the interviews students would state that the ball that was ahead would have to have the greater acceleration.

Many of the difficulties students have with Acceleration Comparison Task 1 seem to be due to confusion between the concepts of velocity and acceleration. The next four sets of interview excerpts demonstrate a failure to discriminate between these concepts. In the first two sets, the concept of velocity is undifferentiated and largely intuitive. The judgment made by each student is almost entirely perceptual. In the second two sets, the procedures seem to be based on a sense of average velocity.

Procedure 3—

precourse

I: How would we compare their accelerations? *(Balls are released.)*

S: Towards the end, they were going about the same rate.

(Demonstration is repeated.) They would eventually reach the same acceleration.

I: How do you know?

S: They had the same speed at the bottom, approximately.

postcourse

I: What does the word *acceleration* mean?

S: It's the change in velocity over a certain period.

I: Do you think, when they go in the tunnel, they have the same velocity?

S: They seemed to.

I: And how do they start out?

S: It seems one would have a greater velocity, initially.

I: Can you use that to compare their accelerations?

S: They had to be equal.

The pair of excerpts above illustrates a procedure in which students attempt to compare accelerations on the basis of final speeds alone.

The student quoted is from the self-paced section of the noncalculus general physics course. The definition given of acceleration as "the change in velocity over a certain period" might well have been accepted as indicative of understanding if the discussion had ended with that statement. As a matter of fact, the student completed the requirements of the course quite satisfactorily and received a top grade—but the fundamental conception has not changed. This student thinks of acceleration as a change in velocity that occurs as time passes, but his use of the word "over" does not imply division. He does not attempt to compare the accelerations in terms of the two quotients obtained by dividing the changes in velocities by the time intervals. Instead, he concludes that the accelerations are equal because the final speeds are the same.

Procedure 4—

precourse

I: If we're comparing two balls and we decide one has a bigger acceleration, what would that mean?

S: It speeds up in a shorter time.

I: Let's see if we can decide which has a larger acceleration or whether they're the same. *(Balls are released.)*

S: They came together. So A had a larger acceleration. It had to because it had to catch up with B which was ahead of it.

I: We would like to decide whether they have the same or different accelerations. *(Balls are released.)*

S: Ball A would have the greater acceleration because in order for A to catch up it would have to go faster.

These excerpts display a procedure frequently used to compare accelerations that is based on the belief that catching up (in the sense of gaining on) means having a greater acceleration. Catching up, of course, merely indicates nonzero relative velocity. In Acceleration Comparison Task 1, ball A catches up to ball B; yet, ball A has a smaller acceleration. The student quoted above completed the calculus physics course in the top half of the class, according to the final course grades, as did all the other students quoted in the rest of this section.

Confusion between the concepts of velocity and acceleration is also demonstrated by students who form an overall impression of the velocities of the balls over the entire motion. These students never focus their attention on separate parts. The two sets of interviews that follow are illustrative of this way of proceeding.

Procedure 5—

precourse

I: Let's see whether these two balls have the same acceleration or different. *(Balls are released.)*

S: It looks like ball A would have a faster acceleration.

I: How are you thinking?

S: Ball A covered more distance in the same amount of time.

postcourse

I: In terms of speed and things like that, how would you describe acceleration?

S: How much the speed changes for any amount of time.

I: Let's see if you can apply that definition to this situation and tell which of these balls has the larger acceleration, or if they have the same acceleration. *(Balls are released.)*

S: Ball A covered more distance, so it must have covered the distance faster to get to the same point at the same time. I would say that A had the greater acceleration.

This student from the calculus physics course assumes that covering a greater distance in the same time means having a greater acceleration. Hence, ball A would have the greater acceleration. In the demonstration, from the instant ball B is released to the instant both balls enter the tunnel, ball A travels a greater distance than ball B. Ball A, however, actually has a smaller acceleration.

Procedure 6 —

precourse

I: How could we argue about their accelerations? They both start from rest and they both reach the same final speed.

S: Well, ball A has a little more time to reach that certain amount of speed. It rolls a little bit before [ball B] is triggered. So it has more distance and more time. Ball B has less distance and less time. And they both reach the same speed.

I: How would the accelerations compare?

S: They might be equal.

postcourse

I: Ball A starts from rest, and reaches some velocity up there. Ball B also starts from rest, and reaches the same final velocity there, right?

S: Right. In less time, and in less distance. So maybe they're proportional, or something, which would make them equal.

I: Ball B reaches that velocity in a certain amount of time, but ball A takes a somewhat greater time to do that.

S: And also a greater distance. I think that the accelerations would be the same.

The calculus physics student quoted in these interviews compared the distances traveled by the balls in a different way and arrived at a different conclusion about the accelerations. This student reasoned that taking a longer time to cover a greater distance can compensate for the greater distance. Thus, even though ball A travels a greater distance than ball B (from its release point to the tunnel), since ball A takes longer, it may have the same acceleration as ball B.

As the foregoing illustrations demonstrate, students sometimes apply procedures that may be adequate for comparing instantaneous or average velocities but that are not adequate for comparing accelerations. To make the correct choice of the ball with the greater acceleration (other than by guessing), a student has to be able to distinguish between the concepts of velocity and acceleration to the extent of being able to identify and compare instantaneous velocities at different times.

Procedure 7 —

S: *(Balls are released.)* B actually had a faster acceleration than A because it catches up to A as far as velocity goes. A didn't slow down but it came to the same speed that B did and B had a shorter distance.

This excerpt from a postcourse interview is reminiscent of Galileo's problem of determining which definition of acceleration to choose: a change in velocity per unit distance or a change in velocity per unit time. Here, a calculus physics student does not discriminate between these two possibilities and associates the change in velocity of the balls with the distance they travel rather than with the elapsed time. Although this student arrives at the correct solution to the task, he does not consider a time interval in his response.

Procedure 8—

S: Ball B's speed increased faster than ball A's.

I: Why do you think that was the case?

S: Because ball A gained position on ball B at a greater rate on the upper part of the track than it did on the bottom part, which I would interpret as the velocities coming closer together. . . . Ball B would have a greater acceleration.

This student from the calculus physics course also makes the correct choice in this postcourse interview. He argues that since ball B has a considerably smaller velocity than ball A at the beginning, but nearly the same velocity later on, ball B must have a greater acceleration. This particular response may or may not represent a conceptual difficulty. It is included here because the student does not make explicit reference to the time interval or compare the accelerations as ratios.

Summary of Student Procedures: The various procedures used by students as they attempted to compare accelerations on Acceleration Comparison Task 1 are listed in Table 2. The first six procedures in the table lead to an incorrect solution. Although the seventh yields the right response, the type of reasoning involved is inadequate. The eighth procedure also results in the correct response but does not explicitly involve the time interval. We have placed this procedure at a lower level than procedures 9 and 10 in which the corresponding time interval is identified.

Analysis of Results. We can present the data from the administration of Acceleration Comparison Task 1 in a semiquantitative form similar to the one used for velocity. Numerical scoring of a student's performance on the task is based on an assessment of overall quality according to the criteria described below.

The performance of students on Acceleration Comparison Task 1 is measured on a three-point scale of 0, 1, or 2. A score of 2 means that the

	Procedure	Interpretation of procedure
1. Same slope	Balls have the same acceleration because the slopes are the same	Nonkinematical approach
2. Final position	Balls have the same or different accelerations depending on their relative final positions	Confusion between position and acceleration
3. Final speed	Balls have the same acceleration because their final speeds are the same	Confusion between velocity and acceleration
4. Catching up	Ball A has a greater acceleration because it is catching up to (gaining on) ball B	
5. Greater Δs; same Δt	Ball A has a greater acceleration because it covers a greater distance than ball B in the same time	
6. Greater Δs; greater Δt	Balls may have the same acceleration because ball A covers a greater distance than ball B in the same time	
7. Same Δv; smaller Δs	Ball B has a greater acceleration because its velocity changes by the same amount as the velocity of ball A but in a shorter distance	Discrimination between velocity and change in velocity but neglect of corresponding time interval
8. Velocity catches up	Ball B has a greater acceleration because its velocity catches up to the velocity of ball A and thus changes by a greater amount	
9. Greater Δv; same Δt	Ball B has a greater acceleration because its velocity changes by a greater amount than the velocity of ball A in the same time	Qualitative understanding of acceleration as the ratio $\Delta v / \Delta t$
10. Same Δv; smaller Δt	Ball B has a greater acceleration because its velocity changes by the same amount as the velocity of ball A in a shorter time	

124

TABLE 3
Scoring Criteria for Acceleration Comparison Task 1

Score	Criteria
0	Student makes little or no use of instantaneous velocities at different times. Student persists throughout the interview in comparing motions in terms of the slope of the incline, the final relative positions, the final speeds, or the phenomenon of catching up.
1	Student compares instantaneous velocities at different times. Student may also compare changes in velocity but does not make explicit use of the time interval.
2	Student considers the ratio $\Delta v/\Delta t$, compares Δv's for equal Δt, or compares Δt's for equal Δv. Student may display inadequate procedures during the interview but successfully resolves difficulties with limited, guided questioning by the interviewer.

student, with limited guidance by the interviewer, is able to make a qualitative comparison of ratios. A score of 1 indicates that the student has compared instantaneous velocities at different times but never refers to the specific time intervals during which the changes in velocity occur. When after a considerable amount of questioning by the interviewer, the student still has made no attempt to use instantaneous velocities at different times for comparing accelerations, a score of 0 is assigned. The specific scoring criteria are shown in Table 3.

The various procedure categories listed in Table 2 were established after scoring of the interviews had been completed. Often during an interview a student would attempt to use a particular procedure, find it unsatisfactory, and then attempt to reason in some other way. Thus, a number of different strategies representing varying degrees of commitment and resulting in varying degrees of success are invoked by a student during an interview. The procedures used by a student do not determine the score. Rather, the score reflects the quality of the student's overall performance, especially her or his final analysis. Students who received a 0 usually relied on procedures near the beginning of Table 2, students who received a 1 generally used procedures near the middle of the table, and those who received a 2 displayed procedures near the end of the table.

Acceleration Comparison Task 1 was included in precourse interviews conducted with individuals from various introductory courses. The groups represented were in-service elementary school teachers, academically disadvantaged students, and calculus physics students. Since the non-calculus general physics students in the self-paced course were presented with a different form of Acceleration Comparison Task 1, data from their precourse interviews have not been presented here. The noncalculus general physics students in the lecture course were not given precourse interviews.

Acceleration Comparison Task 1 was also used in postcourse interviews. It was not always possible to interview the same students before and after instruction for the reasons previously mentioned in the discussion of the speed comparison tasks.

Prior to instruction, between 5 and 40 percent of the students received a score of at least 1 on Acceleration Comparison Task 1. These students were able to discriminate between velocity and acceleration to the extent of being able to identify and compare instantaneous velocities at different times. As might be expected, the academically disadvantaged course had the lowest percentage of students (5 percent) able to meet this criterion before instruction. Approximately 40 percent of the calculus physics students met this criterion. In all introductory-level populations studied, at least three-fifths of the students confused the concepts of velocity and acceleration in precourse interviews.

After instruction, between 35 and 70 percent of the students received a score of at least 1 on Acceleration Comparison Task 1. The percentage from the academically disadvantaged class had risen to 65 percent. *This was comparable to the postinstructional performance of the calculus physics students.* The percentages for both these groups exceeded those for the noncalculus physics students after instruction. In the introductory-level populations studied, about one-third of the students still confused the concepts of velocity and acceleration during postcourse interviews.

A DISCUSSION OF THE RESULTS

The Concept of Instantaneous Velocity

One of the most interesting results to emerge from analysis of the individual demonstration interviews may not be immediately apparent from the data. In both pre- and postcourse interviews, failure on the speed comparison tasks is almost invariably due to improper use of a position criterion to determine relative velocity. *Although the students who are unsuccessful can generally give an acceptable definition for velocity, they do not understand the concept well enough to be able to determine a procedure they can use in a real physical situation for deciding if and when two objects have the same speed.* Instead they fall back on the perceptually obvious phenomenon of passing. Some identify being ahead or being behind as being faster or slower. We interpret this use of position to determine relative velocity as an indication of confusion between the concepts of position and speed.

The Concept of Acceleration

The results of this investigation indicate that introductory physics students frequently lack even a qualitative understanding of the concept of acceleration as the ratio $\Delta v / \Delta t$. Performance on Acceleration Comparison Task 1 reveals *a widespread inability to apply this concept in a real physical situation.* This is the case not only before instruction, but very often afterward as well. *Although almost all the students can define acceleration in an apparently acceptable manner after instruction, most cannot use this definition to determine a satisfactory procedure for comparing the accelerations of two moving objects.*

IMPLICATIONS FOR INSTRUCTION

The fact that many preconceptions prove to be remarkably resistant to instruction suggests that some form of active, experiential instruction is necessary to overcome them. Defining concepts in lectures and textbooks and giving examples of their applications in routine problems do not seem to be enough to bring about conceptual understanding in many students. Specific difficulties need to be directly addressed.

One of the most encouraging findings of the study is the great improvement in performance evidenced in postcourse interviews with students from the academically disadvantaged class. A large number of students in this group began the course unable to discriminate between speed and position, and virtually all were unable to distinguish velocity from acceleration. After instruction, however, almost all could separate the concepts of position and velocity. They also demonstrated a qualitative understanding of acceleration as a ratio that matched that of the students from the calculus physics class.

We are convinced that the type of instruction received by the academically disadvantaged students contributed to this achievement. In their course,[15] concept formation was given special attention, along with the development of scientific reasoning. Confusion among related but different concepts was directly confronted. In laboratory exercises, in class discussions, in individual dialogues with the staff, and on course examinations, the students were presented with situations designed to help them distinguish various kinematic concepts from one another and apply these concepts to the motion of real objects.

For some students the acquisition of physical concepts appears to depend strongly upon the establishment of satisfactory connections between these new concepts and the proto-concepts with which the students are already familiar. In fact, among the academically disadvantaged students such connections seemed to be crucial in order for new concepts to take on meaning. We have found that a conscious effort is necessary to help students relate physical concepts to their experience. These connections often need to be made explicitly.

Although students from the calculus and noncalculus physics courses generally had a more extensive technical vocabulary to describe motion than the academically disadvantaged students, their use of these terms was frequently inappropriate. We found that even students who received good grades in standard courses often had difficulty distinguishing among related concepts and applying them properly to simple motions of real objects.

It has been our experience that deeply seated preconceptions cannot be changed quickly or easily. Unlike the usual situation in an introductory course in which kinematics is covered in a few days, the study of this topic in the EOP course lasted several weeks. Although better prepared students may not require the same amount of time and effort, we are convinced that they, too, would benefit from an increased emphasis on the development of the basic kinematic concepts.

Although the research described here was carried out with college students, the results obtained by Minstrell have established that the same difficulties occur among high school students. In Chapter 9 he suggests several instructional strategies that he has found useful for promoting conceptual understanding among high school students.[16] Some of these same procedures have proved equally effective at the college level.

CONCLUSION

This primarily descriptive study of student understanding of the concepts of velocity and acceleration has yielded some new insights into how students think about motion. The criterion for understanding used in this investigation is the ability to apply these concepts successfully to the interpretation of simple motions of real objects. Several types of conceptual difficulties have been identified through individual demonstration interviews. These findings have been confirmed and pursued further by other methods of investigation not described here.

We feel that there is a need for systematic investigations of student understanding in other areas of physics, and in other sciences as well. The number of major conceptual difficulties identified in this study proved to be relatively small and not unique to any single individual, but they affect concepts central to the understanding of physics. We believe a similar situation prevails for topics other than kinematics.

The research described here has provided guidance for the design of a curriculum to address specific conceptual difficulties encountered in the study of kinematics.[17] It is vital that information obtained from empirical studies of student thinking be used to develop instructional materials that address student difficulties as they *are* and not as we as educators imagine them to be.[18]

Conceptual Development Research in the Natural Setting of a Secondary School Science Classroom

James Minstrell

Perhaps nowhere more than in the teaching of physical science ideas do we encounter so great a mismatch between students' intuitive explanations and scientists' explanations of scientific phenomena. These conflicts first become very apparent at the high school level of instruction. For example, consider the situation of a small-wheeled cart rolling across a horizontal, smooth table top. When asked to explain the motion of this cart, students will often suggest that the cart has been given motion by the push of the hand (or whatever), but that the push is "used up" in moving across the table. When asked to describe the various pushes (or forces) that would have to act to keep the cart moving at a constant speed in a straight line, many students will readily hypothesize the necessity of a forward force bigger than the backward force resisting motion. Even when confronted with a nearly frictionless situation, they argue that the hand force must still be with the cart, even after the cart left the hand, or it would come to a stop. When an object does come to a stop, the explanation often includes the "wearing down" of the hand force by friction and/or by the weight of the object.

In some instances it appears that the students are conceiving of force as an action necessary to keep in motion an object whose "natural" tendency is to be at rest. In other cases, students seem to be conceiving of force as an object's property, like inertia, impetus, momentum, or vis viva (the historical precursor of kinetic energy).

The example serves to describe two general sorts of phenomena that are only recently receiving significant research attention. First, students appear to come to the instructional setting with some sort of conceptual structure already in their heads that prompts them to organize the phenomena in their own way. If you ask them to make a prediction about what will happen during an interaction between physical objects, students almost always have some idea. They usually can and do make a prediction. The frustrating thing for teachers lies in the fact that these "natural" explanations hang on in spite of instruction. While students may well change in their ability to answer questions that are similar to those used as examples during instruction, they often cannot apply the physics ideas appropriately in a new context. With the exception of attempting to use different words or formulas, the students answer a novel context question in the same way that they answered it prior to instruction.

One reason that students fail to change their basic premonitioning could be that most of our present curriculum and instructional strategies do not take into account the initial conceptions with which students come to our science classes. We tend to try to superimpose our "formulaic" products from science over the students' existing way of organizing the world. As long as there is no mismatch between what the students already know and what we teach them, things go smoothly. But when there is a mismatch, students memorize the formalisms of science like some kind of foreign language. Science class becomes some sort of foreign culture that they visit once a day, rather than a place where they get together to discuss and investigate the phenomena of the world around them and where they have an opportunity to develop a more consistent organization of the phenomena.

The purposes of this chapter are threefold. First, the reader will become more aware of the phenomenon of alternative conceptions and the methods of investigation in conceptual understanding research to which classroom teachers could make important contributions. In Chapter 8, McDermott has developed the basic rationale and applied it to college science. In this chapter, I will extend the rationale and apply it to high school physics (to work I am doing in my own classroom at Mercer Island High School) and briefly compare my methods and results with those of McDermott and others at the University of Washington. Second, the reader will be asked to consider the implications of student preconceptions for instructional strategies and curriculum development. I will describe what I believe to be some implications for teaching science that result from conceptual understanding research. Finally, I challenge every classroom teacher of physics or physical science to join in the efforts of researchers to identify and understand students' alternative conceptions and/or to join in the efforts of developers to incorporate these new findings into curriculum design and instruction for the 1980's. The task is not easy—but it is intriguing.

130

METHODS AND RESULTS OF INVESTIGATING ALTERNATIVE CONCEPTIONS

Context for This Research [1]

The initiative for my research on alternative conceptions grew out of the frustrations of my own teaching experiences, particularly the frustration that my very rational (to me) instructional activities were less effective than I desired. Through listening to my students, it became clear that it was not merely a case of their not being bright enough to understand my imposed lessons; many students had a consistent conceptual structure for understanding the world, but their conceptual understanding was an alternative to the physicist's point of view. In many cases, their alternative conceptions were naive, but in general, they served the students in their day-to-day work.

The research described here was conducted over a two-year period in the naturalistic setting of two physics-related classes at Mercer Island High School. Nearly all of the 55 students had professional parents and have gone or will go to college after graduation. The average high school cumulative grade point for these students was close to 3.5 on a 4.0 scale. Many of these students had strong science and mathematics backgrounds. Nevertheless, "even" these people who were bright and aware exhibited evidence of harboring conceptions that were fundamentally at odds with what I was trying to teach them.

Data for this study were gathered in the context of the activities of these physics classes. Typically, prior to studying a new unit, a pre-instruction quiz was administered to determine the extent to which students were using alternative conceptions. They were usually paper-and-pencil quizzes and included questions designed to be sensitive to alternative conceptual structures. Some questions were adapted from those used by other researchers (see the "References" to Chapter 8 for some useful sources). Other questions were developed based on difficulties identified in my classes in earlier years. Other data or alternative initial conceptions came from tape recordings of large- or small-group discussions within the classes. I monitored the tenacity of the alternative conceptions, as would any teacher, through paying careful attention to what they said in discussions, wrote in lab reports, and did on tests that I constructed. The way the tests were designed is an important feature of the investigation.

Categorized first into correct and incorrect sets, the explanations were then examined to see what conceptions were being employed and how they meshed with the formal system of physical ideas.

The accuracy in determining whether an individual's thinking was governed by an alternative conception depended upon several factors. Determination with a single pre-instruction test question depended on the quality of wording of the question and on the individual student's willingness and ability to clarify his or her thinking. Evidence accumulated

131

from answers to several questions, sometimes given on different tests or offered in different contexts, helped to establish reliability and validity of the measures. In some cases, some individual students were orally questioned regarding their answers in order to probe their thinking.

Investigations in Kinematics

Prior to any formal instruction in the concepts of motion, the two physics classes participated in a demonstration adapted from the Speed Comparison Task 1 (passing twice) developed by Trowbridge and McDermott[2] and described in detail in Chapter 8.

The students were asked to describe separately the motion of each of the two balls. They observed one ball moving at uniform speed and one slowing down. I then asked whether the two balls *ever* were traveling at the same speed and how they knew whether or not they were moving at the same speed. Then I demonstrated the motions of the two balls in slow motion by moving the balls along the track by hand.

At least 30 percent of the students were clearly confusing "same speed" with "same position." They would make statements such as these:

> When A passes B, at that one point they have the same speed.

> At two points they travel at the same speed, at about 15 cm and at about 75 cm mark. You know this because they parallel each other for a second before one of them passes the other.

About 50 percent of the students quizzed were able to correctly identify one time somewhere in the middle of the journey when the two balls were traveling at the same speed. About half of these students used a perceptual-based argument that clearly indicated that they did *not* confuse speed with position; for example:

> . . . where the *distance* between balls stops changing.

> . . . about 3/7 of the way along the track for ball A and about 4/7 along for ball B. At that time they reach the same speed.

The other half of these students used a logical argument to justify their answer that there was a time when the two balls had the same velocity. Typical answers given by these students include the following:

> Since ball B is moving faster than ball A at the beginning and then at the end is going slower, then some point during its deceleration it reached the speed of A.

> At one instant, one is moving faster but slows down. The other moves with almost uniform speed, so at one point while the other changes from faster to slower speed, they must have the same speed.

These logical arguments are based on assumptions about the continuity of velocity from faster to slower and the deduction of "some" point where the two velocities must be the same. These arguments leave room for the possibility that a student still might fall into the speed–position confusion in another context. Indeed, a few of these students did exhibit the confusion in other contexts later in the course. The point here for the teacher of introductory physics students is that these students were able to answer the questions correctly on the basis of a nice logical argument, but, nevertheless, several of them were still harboring the confusion that "same position" indicates "same speed."

A few of the students using logical arguments least were not harboring the speed–position confusion, possibly because of the pattern of observations they made:

> If the two speeds never equalled each other at any point, then the one ahead would always remain in front of the other. This is not so, as they each passed the other once.

Of the remaining 20 percent, some demonstrated conceptual reasoning troubles that apparently blocked them from identifying any possible instants when the speeds of balls A and B were the same. Others gave unclear answers and sufficient vague justifications that their conceptions could not be known from this group-administered demonstration quiz.

It is important to note that the results of investigations into concept understanding at the high-school level are very similar to the results reported in Chapter 8 at the university level. At least 30 percent of the introductory physics students confuse "same position" with "same speed." That is especially interesting in light of the fact that the methods for studying the conceptions were quite different; McDermott's data were gathered in a clinical interview setting while the data here resulted from research done in the natural setting of classroom experiences.

There are tradeoffs in the methodology. In the clinical interview, one can be relatively more certain about the thinking of each student interviewed, but the procedure is costly in terms of time. In the classroom demonstration format, one is less certain of the thinking of each individual, but one can obtain approximate percentages of the group that exhibit certain difficulties in a relatively short period of time. Knowledge of the thinking of individuals in the group can be obtained by accumulating other samples of data on the same student (e.g., other paper-and-pencil tests, homework, or verbal statements made during class discussions or individual conversations). Generally, if a student has difficulty describing a way to determine when the balls move at the same speed on the demonstration quiz, he or she will have difficulties separating speed from other concepts in other contexts.

As might be expected, the format in which the problem is posed affects the results. Generally, the more abstract the context, the greater the incidence of confusion between conceptual ideas. A graphing context, for example, will usually elicit more errors than a "live" demonstration. Recently my students were considering three velocity graphs, each of which depicted the motion of a car that started with an initial velocity of zero and after four seconds had a final velocity of 8 m/sec. Graph A was concave down, B was concave up, and C was a straight diagonal. (See Figure 1.) When asked which car went the farthest, two students in one class argued that A and B traveled equally far and that both traveled farther than C because "they had to travel along a curved path and C traveled in a straight line between the same two points." Apparently these students were perceiving the graph of the object's motion as the actual path of the object. This difficulty in extracting correct meanings from the graphs was discouraging in the face of the fact that these two students were doing well in their fourth year of high school mathematics; one was taking calculus. It appears that the connections between the graphs and the physical events are not being made.

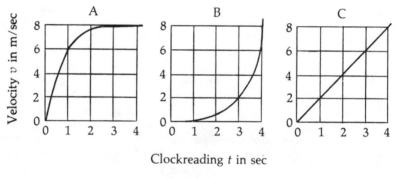

Clockreading t in sec

FIGURE 1
THREE VELOCITY GRAPHS

Investigations into the Nature of Gravity

In past years, there were strong indications among my students that their conceptions of gravity and its effects were different from mine. Occasionally, for the sake of argument, I have asked my students to consider what the world would be like without the surrounding air. Usually someone would suggest that, without air, objects would drift up off of floors, tables, etc. Could they have been implying that air pressure is responsible for gravity? In other arguments, we considered dropping and/or throwing projectiles of various masses, and many students predicted that heavier objects would take less time to fall.

As a result, prior to studying the nature of gravity and its effects, I administered a pre-instruction quiz. Students were given a sheet of paper with diagrams and a few words to represent questions relating to gravity (Figure 2). Because the general descriptions of the problem situations were

134

lengthy, they were delivered orally. Instructions for problem 1 were para-phrases of the following:

The diagram represents an experiment that is done in two different ways. First, the picture on the left represents a sturdy frame from which a spring scale is suspended. On the other end of the spring scale a weight is hung. The scale reads 10.0 pounds with the weight attached as it is.

Secondly, on the right, we repeat the same experiment except that the entire experiment will be done under a huge, strong, air-tight glass dome. Then, we use a huge vacuum pump to take all the air we can out of the space under the glass dome.

In the answer space provided, indicate approximately what the scale would be expected to read. Beneath that, explain why you believe the scale would have the reading you predicted.

Twenty percent of my students predicted that the scale would read substantially less than 10 pounds. Typical justifications included the following:

Because without air molecules in the atmosphere there would be a void and in a void there is no weight.

In the bell jar, it doesn't weigh anything since all the air has been taken out which changes the gravity.

Because in a vacuum things weigh almost nothing, like spacemen.

Because there won't be any more gravity.

For some students, air pressure causes gravity, so when you use the vacuum pump to evacuate the air, gravity is gone. Another possible con-ception is that both air and gravity are within the glass jar, and the vacuum pump takes out both the air molecules and whatever "stuff" grav-ity is made of.

One might ask how this sort of conception could have been for-mulated. It appears that the students' limited knowledge of the experiences of astronauts may be partially responsible; several of the justifications included references to astronauts or spacemen. An association has appar-ently developed between an airless environment and lack of gravity. On the moon, the effect of gravity is less and there is not sufficient air to breathe. In outer space, there is no air and things just seem to "float" around.

Problem 2 on the pre-instruction quiz was posed orally as follows:

In normal air near the surface of the earth, a 1-kg sphere is dropped from a certain distance and falls for 1.0 sec before hitting the ground.

In the second diagram, a sphere of the same size but with a mass of 5 kg is dropped from the same height. Below the diagram indicate roughly how much time it will take this object to fall.

About 25 percent of the students suggested that the mass of the object greatly affected the time of fall; most suggested that an object five times as

Nature of Gravity and Its Effects

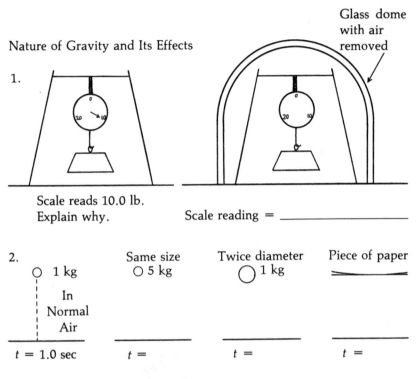

1.

Glass dome with air removed

Scale reads 10.0 lb.
Explain why.

Scale reading = _____

2.

| | Same size | Twice diameter | Piece of paper |
| O 1 kg | O 5 kg | O 1 kg | |

In
Normal
Air

$t = 1.0$ sec $t =$ _____ $t =$ _____ $t =$ _____

Briefly explain how you arrived at your answers.

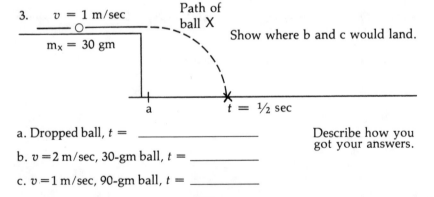

3. $v = 1$ m/sec

Path of
ball X

$m_x = 30$ gm

Show where b and c would land.

a

$t = \frac{1}{2}$ sec

a. Dropped ball, $t =$ _____   Describe how you
                                            got your answers.
b. $v = 2$ m/sec, 30-gm ball, $t =$ _____

c. $v = 1$ m/sec, 90-gm ball, $t =$ _____

FIGURE 2
SHEET ON WHICH STUDENTS ANSWERED QUESTIONS POSED ORALLY
TO THE CLASS

heavy would take 1/5 the time to fall. A few students suggested that the heavy object would fall faster but answered with "5 sec." The arithmetic on their papers suggested that they also had trouble with proportional reasoning where an inverse proportion was involved.

In the third diagram, a sphere with a mass of 1 kg but with twice the diameter was dropped from the same height. Below the diagram indicate roughly how much time it would take the object to fall.

Over 20 percent of the students suggested that the time of fall would be greatly affected by the change in diameter. Most of these students argued that doubling the diameter would double the time of fall.

In the fourth diagram, a sheet of notebook paper is held horizontally and dropped from the same height. Below the diagram indicate roughly how much time it would take for the object to reach the ground. Explain your reasoning.

Nearly all of the students concluded that the paper would take much more time to reach the floor. Most arguments involved a large air resistance, but an occasional student argued that if its weight was 1/20 of 1 kg, then it would take 20 times as much time to reach the floor. Also, an occasional student argued that the paper would take just as long as the 1-kg sphere because he/she remembered from a *film* that "all things fall at the same rate." (This is an example where such a statement clearly conflicts with their experience.)

Instructions for problem 3 were paraphrases of the following:

A 30-gm ball is rolled horizontally along a high table top. The ball moves with a velocity of 1 m/sec and takes 1/2 sec from the time it leaves the *edge* of the table until it hits the floor. It hits the floor at point X.

(a) Suppose a 30-gm ball is dropped from the edge of the table. It will fall straight down and land at the base of the table. How much time will a ball take to go from the edge of the table to the floor?

(b) Suppose a 30-gm ball with a velocity of 2 m/sec is rolled off the table. How long will ball b take to go from the edge of the table to the floor? Where will it land?

(c) Suppose now a 90-gm ball with a velocity of only 1 m/sec is rolled off the table. How long will ball c take to go from the edge of the table to the floor? Where will it land?

Describe how you determined your answers.

In this context of projectilelike motion, about 50 percent of the students made predictions for the time of falling that were not consistent with the physicist's view: they suggested that the time of falling is affected greatly by differences in mass. Of particular note is the failure of students to be consistent. Over half of the students who inferred that mass did have an effect on the time of fall in the projectile situation suggested that mass did not have an effect in the dropped ball situations of problem 2.

In reading these students' justifications for problems 2 and 3, it appears that many students' knowledge of comparison times for dropped objects had' been influenced by their educational experience. Typical answers for problem 2 included references to statements made by former teachers or films depicting Galileo dropping balls from the tower. Many of these same students then answered problem 3 as though mass affected time. These students apparently learned or memorized a result under certain conditions but had not yet sufficiently understood the result to be able to generalize to a context where there were two components of motion.

Another major alternative conception surfacing out of problem 3 is that mass affects the distance that the ball will travel horizontally. About 70 percent of the students predicted that the 90-gm ball would not travel as far as the 30-gm ball, given that they are launched with the same velocity. Only about 20 percent of all the students answering problem 3 correctly concluded that neither mass nor velocity greatly affects time of fall and that mass does not greatly affect how far the object travels.

With varying results like those for problems 2 and 3, how does one decide which is the better representation of the students' thinking? In my assessment of the conceptual understanding, if an individual student uses an alternative conception in any of several contexts, I then assume that either the student holds that belief as part of his or her conceptual framework or that he/she is in transition between conceptual frameworks. Thus, with information from both problems 2 and 3, I gain confidence that a student does not hold an alternative conception by finding that he or she apparently doesn't use that conception in either context. Second, if I only have one measure, I would choose problem 3 because although it is a common experience, it is not as likely to have a "taught" result. After instruction, the greater the number of contexts posed in which the student does not use a particular alternative conception, the more confident I feel in claiming that the alternative conception is *not* now part of that student's conceptual framework.

IMPLICATIONS FOR INSTRUCTION
TOWARD EFFECTING CONCEPTUAL DEVELOPMENT

An Awareness of Initial Conceptions

It appears that when the teacher is aware of students' initial conceptions and incorporates that knowledge into instruction, development of appropriate understanding is enhanced.

Another important instructional strategy that seems to foster change in concept understanding is to force the student to state his or her initial conception and then relate it to other concepts. To help the student apply his or her beginning conception, the teacher needs to know what that student's present thinking is like. One specific technique is to ask students for their predictions and then ask them how they arrived at those predictions. Another technique is to ask students to explain an observation they have made. If students are familiar with the physical situation being discussed, and if students are allowed to answer fully without fear of ridicule, they usually can and will make a prediction or offer their explanation of an observation. Note that a feeling of mutual trust and respect can help greatly.

Generally, in my classes, the students who are most prepared to note discrepancies with their existing conceptions and to seek conceptions more consistent with all the related phenomena have participated most candidly in a pre-instruction, open demonstration–discussion or even the semi-private experience of a pre-instruction quiz.

Sometimes the simple public experience of participating in a quick poll or vote will heighten the students' readiness for conceptual change. For example, prior to a demonstration–discussion on forces on static objects, I asked students to consider a book on a table. I asked how many believed that the table exerted an upward force on the book and how many believed that the table did not exert an upward force. Approximately 50 percent of the students "voted" for each alternative. There were bright, articulate students on both sides of the issue, and the stage was set for a lively discussion of whether it was useful to think of tables, springs, hands, trampolines, etc., as exerting forces. One should be careful in using this technique that students "voting" for the "more naive" conception do not develop feelings of inadequacy. I use this technique only when I am reasonably confident that at least 40 percent of the class will support the "more naive" conception.

Sense Experience

In the past, *sense experience* has been narrowly defined as "doing laboratory experiments." As Bates[3] concluded in his summary of research regarding the effects of laboratory experience in the classroom, students do not necessarily come out of the course with greater knowledge of concepts. From my experiences as a university student and from my observations of high school students doing laboratory experiments, students are generally focusing on following the procedural "recipe" for gathering data, analyzing results, and answering the necessary questions at the end. That is, much of what we've called hands-on laboratory experience has resulted in a procedural activity or a skill-based activity rather than an activity to build knowledge and awareness of phenomena. But when properly cultivated, sense observations from laboratory activities, demonstrations, or discussions of past personal experience *can* provide the knowledge and awareness of phenomena that will foster change in *conceptual* understanding.

Many times sense experience by itself will convince students that their present conception of the world is not adequate. Consider the aforementioned conception that the time of fall for two spheres is greatly affected by the mass of the spheres; place a wooden sphere and a metal sphere on a book held horizontally above the floor and jerk the book downward and out from under the spheres, releasing them at the same time. The virtually single sound of them hitting the floor at the same time is usually sufficient to convince students that a conception of "fall rate depending upon mass" is inadequate.

Another example of a sense experience's affecting students' conceptual structure comes from astronomy. More than one third of my students come in believing that the earth's shadow is responsible for the dark portion of the phase of the moon. However, most of these students realize that their conception is not adequate when they are confronted with the observation that the moon appears in the sky near the sun when the moon is in its "thin crescent" phase, implying that the sun–earth–moon angle (with earth at the vertex) is much less than 90 degrees; the earth's-shadow argument would be valid only if the moon were partially behind the earth relative to the sun, an angle of nearly 180 degrees. The observation may be made by the student or shared by another student, but the important factor is that the student is confronted with the observed fact and encouraged to make sense of it.

To reconstruct a "better" conceptual framework after discovering the inadequacy of the present one seems to involve accumulating several related observations. For example, to build a conception to account for phases of the moon often takes an accumulation of observations of sun–earth–moon angles associated with their respective phases, and it may also include observations of "phases" created by illuminating a sphere with a far-away, bright light source.

Many times an analogue with students' own common experience will help them realize the inadequacy of their present concept. Even those students who answered the rolling balls demonstration question with, "The balls are going the same speed when they are next to each other" (points of passing) recognize a conflict when it is pointed out: "That's like saying a car in the passing lane is traveling just as fast as a slow car in the outside lane (or a stopped car at the side of the road) just because at that time they are next to each other on the highway" (an argument suggested by some of my students).

Interaction Between Sense Experience and Rational Argument

Construction of a new conceptual framework seems to require an interrelationship between sense experience and rational argument. For example, consider the 50 percent of my students who believed that a table could not exert an upward force on a book on its surface. After viewing several demonstrated conditions of static objects (a book on a table, a book on an outstretched hand, a book suspended from a spring, and a book on a perceptably bendable table) and after discussing the similarities and differences in these experiences, students began seeking a rationally consistent argument for explaining the common effect, the "at rest" condition of the several different observed situations.

Another example of the apparent interaction between sense experience and rational argument occurred during the discussion following the rolling balls demonstration. The following arguments in combination seemed to convince most students that "same position" was not sufficient for identifying a time when the balls were rolling with the "same speed," and they suggested a believable strategy for identifying a time when the balls were moving at the same speed: from common experiences in cars, the one that is passing the other is going faster. One ball passes the other and then is passed by the other. Therefore, there must be some time when they are going the same speed. Returning to the car analogy, the way to identify when your car is traveling the same speed as the car ahead is to note when the distance separating the cars is not increasing or decreasing. So the balls are traveling at the same speed when the distance between the balls is no longer increasing and is not yet decreasing. These were arguments offered by some students; other students seemed convinced in the discussion.

Alternative conceptions do not usually dissolve and go away easily. In many cases, the conceptions held by students represent a consistent organization of phenomena that has served them for years. It is not likely that one lecture, discussion, or laboratory period is going to bring about a new conception in any permanent way. Even with instruction that attends to student misconceptions, it may take many diverse experiences aimed at eliminating that particular misconception before a student completely gives up the old notion. In a context of graphing, many of the students who were seemingly convinced that "same position" did not mean "same speed" inferred that when the graphs of speed of two objects crossed, that meant that the two objects were at the same place. Many repetitions of situations involving the potential speed–position confusion seem to be required before the sense experience and rational argument that make sense in one context can be generalized to virtually any context.

The rational arguments that students use do not always appear to be the formal logical operations identified by Piaget and Inhelder.[4] Often the student starts with likenesses and differences between observed situations and searches for a consistent rationale for organizing the phenomena. For example, when considering the forces necessary to keep an object in the static condition, students observed several diverse static situations, noted the similarity of effect, and argued for a "consistent" way to explain the phenomena. Some students were able to arrive at a new conceptual organization by employing rational operations on a relatively small base of sensory experience, while others required much more sense experience before being able to come to a rationally consistent whole. These rational processes should be encouraged since they are probably primitive versions of the more formal logical operations of controlling and correlating variables.

Appropriate use of questioning can provide an environment in which students' logical reasoning skills are actively involved in organizing phenomena into a meaningful conceptual whole. After students have had ample time and opportunity to become aware of the observations, teachers can ask students to generalize their observations into a consistent pattern, to justify their answers through the use of observations and rational argument (rather than by authority), and to clarify their ideas (to state them another way or to elaborate on them).[5] Teachers who ask a question and then wait while the students consider their possible answers can foster longer, more creative, rationally consistent answers.[6]

The implications for instruction designed to encourage conceptual development are becoming increasingly clear. Teachers need to help students become aware of their initial conceptual thinking. Students' preconceptions can be altered in the direction of accepted scientific thinking through guided sensory experiences. But these sensory experiences may have no effect unless students are encouraged and motivated to consider these new experiences in light of their existing conceptions and to rationally seek a new, more encompassing, and more consistent organization of observations and ideas from which they can answer questions about the natural world. Teacher skills in the appropriate use of questioning and listening techniques are needed to guide students in this personal inquiry process.

CHALLENGE TO THE CLASSROOM TEACHER

Become Knowledgeable of Students' Alternative Conceptions

This chapter, as well as Chapter 8, adds to the growing evidence that students come to our classrooms with preconceived concepts—our students are not the blank slates or empty vessels that many of our teaching strategies have assumed. When we attempt to impose the formalisms of science on our students without taking into account their initial conceptions of the world, they may learn to apply the formulas well enough to get even an A grade in the course, but we may not have altered their conceptual organization of the phenomena in any meaningful way. Often even the skills at applying the formulas "atrophy" within a few months. To be more effective science teachers, we need to know more about our students' precourse conceptions and then base our instruction on the intellectual needs of our learners.

We need to become aware of the results of research on concept understanding. The references for this chapter and Chapter 8 provide a starting point. Published results are often slow to appear, but research reports are usually available directly from the project directors. Both the National Institute of Education (NIE) and the National Science Foundation (NSF) have been funding cognitive process research, and copies of research reports are available from these organizations.

Become Active in Research and Development

Probably the best source of alternative conceptions relevant to your teaching is your class. My own efforts in conceptual development research grew out of the realization that there was more to student lack of understanding than not being bright or not being motivated. At the beginning of your next unit of study, ask your students to tell you what's going on in some natural situation: What causes the light and dark parts of the phases of the moon? Where does the moisture come from that collects on the inside of windows on a cold day? What are the forces (pushes) on a book to keep it at rest? Or in uniform motion? Or uniformly accelerated motion? If you can listen carefully and respectfully to what they say and avoid giving a quick lecture on the subject, you may find that some answers are very well formulated, even though they are conceptually incorrect. Record those answers, share them with colleagues, try to understand the students' thinking processes, and then invent the instructional sequence and activities that begin with the students' conceptions and lead them toward the view of the scientist.

CONCLUSION

This chapter has described methods, results, and implications that stem primarily from one project. The methods used in this project were naturalistic, being conducted in the setting of a high school physics class, but the results were similar to those of university projects incorporating clinical procedures. This study demonstrates that basic research on concept understanding can be conducted effectively in the natural setting of the science classroom.

It is hoped that the results of this project will contribute to the conceptual understanding research data base from which a theory of conceptual development might grow. The actions and statements made by the students are our observations, our "objective reality." From these observations, we infer conceptual structures used by the students to organize the natural world. We infer factors affecting the development of those structures, and we infer implications for curriculum and instruction. The reader is cautioned to keep observations and inferences separate. It always seems a bit ironic to me; here *we* are developing concepts about conceptual development.

The results of conceptual development research will become a major factor in the development of curriculum and instruction for the 1980's. They will help clarify the valuable aspects of various teaching modes: laboratory, lecture, demonstration, discussion, questioning, listening, etc. They may lead to greater clarification of the meaning of science education and its goals and objectives. Finally, and most importantly, they could lead to more efficient, meaningful learning for students.

CHAPTER 10

The Computer and the Teacher

Joseph I. Lipson
Laurette F. Lipson

Teachers have always used technology to facilitate teaching and learning. The age-graded classroom with its desks and blackboards was an impressive innovation in its time and is still serving us well. It creates a stage, an arena for learning in a social context. The technology of printed texts and other printed materials provides information that limits the subject matter authority of the individual teacher ("But, teacher, the textbook says that's not right.") while expanding the knowledge available and the flexibility of instruction. However, today we can break out of the limitations of print. With print we can teach the parts of a radio and the principles of how a radio works, but we have difficulty teaching how to construct and troubleshoot a radio. For these skills we need something more. In the absence of a sophisticated electronics lab, we might try to use slides, tapes, films, videotapes, etc. The newest educational tool, the computer, offers us the opportunity to bring new kinds of experience to our students. While there are many imaginative uses of the computer that we could discuss,[1] we will focus on computer simulations to illustrate the potential of the computer in augmenting the teaching process under the guidance of the classroom teacher.

First, we will describe what we believe every teacher wants to know about computers. We will provide some guidance on how to gain a sense of what is happening and how to become comfortable with computers. Finally, we will explore the future possibilities of the computer and the way in which computers will affect the teacher's role. We will especially consider the possibilities for science and math education.

COMPUTER LITERACY FOR TEACHERS

Steps Toward Literacy

In a 1979 survey Edwards[2] has found that teachers with some exposure to computers feel that they need the following kinds of knowledge:

1. How to use computers in their teaching areas
2. How to increase the use of computers in their schools
3. Where to obtain help in using computers in instruction
4. Where to obtain information about the computer systems and instructional software (i.e., computer programs) available.[3]

There have been, and probably will continue to be, a stream of books and magazine articles written for people who are just beginning to learn about computers.[4] As a result, you should be able to find material in the style and at the level that suits you. For example, the October 1979 issue of *The Practitioner*, a newsletter published by the National Association of Secondary School Principals, is devoted to the theme "MICROCOMPUTERS . . . The Future Is Now." The articles do a good job of introducing key terms and ideas in simple language. The September 1980 issue of *Instructional Innovator* also seems especially useful.

Turning to some specific suggestions, the following is a list of steps to take if you are beginning to think about using computers in your class:

1. Become familiar with the inexpensive computers that are available or are about to become available.
2. Find out what each system looks like and how the parts work.
3. Gain a mastery of basic computer vocabulary such as:
   input device—typewriter, joy stick, touch panel, graphics tablet, etc.
   output device—printer, loudspeaker, television screen, etc.
   memory—means for storing information electronically
   central processing unit—the heart or brain of the computer which carries out logical operations according to a program or a set of very precise instructions.
   Vocabulary is essential. One teacher has remarked, "I want to ask a question, but I don't even know the words for the things I want to ask about."
4. Learn the procedure for starting up a computer lesson on some common microcomputers.
5. Find out what kinds of lessons are available in your field.

6. Gain hands-on experience with a variety of lessons, simulations, and computer games.

7. Acquire an understanding (an ability to describe and explain) of the capabilities and languages of various computer systems (e.g., Radio Shack TRS-80, Apple, Commodore Pet, and others).

8. Build up a file of articles, books, information sources, and bibliographic resources.

Of course, there are a great many useful computer concepts. Initially your grasp of some ideas will be limited. However, teachers can appreciate that as one gains more experience, each concept will acquire deeper and richer meaning. The important thing is to start, to overcome the insecurity you may feel about this new kind of medium. As you begin to read about, talk about, and work with computers, you will find that your vocabulary will grow and your understanding of the terms will become more powerful. Initially, it helps to have someone, preferably another teacher, whom you know and trust act as your guide—someone you can ask questions of without fear of appearing foolish. Increasingly, the marketing representatives of publishers and computer manufacturers will eagerly answer any questions you may have. Members of computer clubs and employees of computer stores are also skilled at answering questions.

As you become comfortable with the idea of what computers can do, you should start to look at the computer as another tool of your trade—like the blackboard, the wall chart, the textbook, or the television set. Keep an inquiring mind and try a variety of new programs at workshops, computer shows, teacher conferences and conventions, computer stores, computer club meetings, etc. Begin to think of the way that *you* want to use the computer and the features that *you* would like to have. Try to analyze possible applications in terms of your own problems. Ease into a working relationship with computer technology. Ask students and parents about their experiences with computers, and keep a file of students, their parents, and other teachers and friends who have relevant knowledge and abilities. A good way to start thinking about computers with your class is to assign the students to analyze how computers are affecting our everyday life (computer checkouts at the supermarket, computers in devices such as microwave ovens and sewing machines, computer banking machines, computer systems for making reservations for travel or entertainment, etc.).

Even if you never develop a computer-based instructional lesson, you can become knowledgeable and able to participate in decisions regarding how the computer will be used. Talk with other teachers; build a notebook–diary of your thoughts and experiences. To enter the "computer culture," you have to do a lot of talking with other people, "visit" computers, and get help in making them do something.

Individual Differences

Several studies have shown that people fall along a continuum with the following two extremes: (1) people who are mainly interested in objects (e.g., machines) and what they can do; and (2) people who are interested in the flow of interpersonal communication. We would propose that the communication or people-oriented individuals might shy away from computers, while the object-oriented among us will eagerly embrace these fascinating toys. Teachers, by and large, are people-oriented and, thus, may have some difficulty in feeling comfortable with the new information machines. What can be done?

We propose the following approach: Those who are people-oriented should be introduced to computers in a group situation that has well-defined roles and a lot of social support for the newcomer. Socially oriented persons should be given assignments to report on to a group (the larger, the better as long as stage fright doesn't take over). They should be provided with role models and rules that give them permission to make mistakes, to feel uncertain while they are gaining their initial knowledge of computers. In other words, socially oriented people should learn about computers in a social setting. It should be made clear that the group is supportive, but that eventual competence is expected. We recommend that there be group activities and assignments in which two or more people act as a team. Teaming encourages conversation, and conversation encourages assimilation of content and processes.

Object- and action-oriented people should be set right to work on individual and team activities that require them to make the computer do tricks and that require them to explain what they did once they got the computer to do something. The challenge need not be too difficult. For example, just learning the procedure for getting a simple program to run on a microcomputer can give the action-oriented person a sense of satisfaction. The action-oriented person gets a sense of effect by making things happen and (hopefully) by understanding how and why they happen. The socially oriented person gets a sense of satisfaction by operating successfully as a member of a social group.

Once you feel you know where you fall on the continuum of social vs. object orientation, you can either select the appropriate kinds of experience from those that are available or begin to organize and negotiate for new opportunities. For example, if a course on computer literacy is being offered, the socially oriented members of the class might be able to negotiate with the instructor for more team projects, more field trips that have a social component, and more class presentation by students. The object-oriented members of the class might request immediate assignments that allow them to roll up their sleeves and begin to make the computer perform.

Better Problem Solving: Bugs and Debugging

Giving a student experience in constructing computer programs can change her or his attitudes toward mistakes. When run the first time, programs almost always have bugs—errors of one kind or another—in them. Even experienced programmers expect them as a natural result of doing something complex. Many times bugs are difficult to locate, but the program will not run until they are all eliminated. Students quickly learn that bugs are common, and that they must be found and removed. No teacher has to keep after them to correct mistakes. Their programs just will not work until the mistakes are corrected. Thus, the incentive to improve a program exists rather naturally in the computer environment. Students feel more responsible for what turns up on the screen since the results are so immediate and changes can be made so quickly. In the world of programming, mistakes become a positive source of information; i.e., they actually lead to an improved product.

We can learn a great deal about students' fundamental approaches to problems by observing what they do when faced with the necessity of debugging programs they have written. Seymour Papert, a mathematician at MIT, has been helping young children learn to program computers themselves and, thus, to develop powerful new ways of thinking. In his book *Mindstorms*[5] Papert illustrates how their rapid mastery of LOGO, a powerful graphic language, brought the children into contact with fundamental ideas from science and mathematics. Over many years he has watched what students do when a program needs debugging, and his observations have implications for the teaching of problem solving generally. What the expert sees as a good program in need of a little fix-up in one or two spots will be treated as a small disaster by some children. If a program does not run, some children erase the whole thing. Instead of debugging, these children erase (which is easy to do on a computer) and start over. Papert argues that such children see a mistake as "wrong," "bad," something to hide. The debugging philosophy, on the other hand, develops a different attitude toward errors. Students who acquire a good debugging strategy develop "fix-it" skills. To them, most programs are not wholly wrong; they need to understand what parts went wrong in order to patch things up. The incentives to improve that are inherent in a computer programming context are powerful—a program will run to completion only when the bugs are out.

Papert thinks it is too bad that computer-aided instruction has simply meant using computers to program children when the greater potential lies in teaching children to program computers. He provides evidence that this latter approach can bring children into a working association with more complex relations in science and develop their mathematical intuition. He illustrates his contentions with some of the models they have developed—and the children in his programs have not been selected because they show any particular aptitude patterns!

148

Programming, coupled with the use of simulation, has been particularly helpful in teaching physics to all types of students at the University of California, Irvine. Alan Bork, a professor of physics and information sciences, has taken the lead in developing a rich array of innovative programs that take advantage of the powerful technology available in micro- and minicomputers. The testing of evaluation procedures and programs will eventually enlarge the current testing conceptions—a much richer array of alternative responses can be made available when the test takes place at a terminal.

Gordon Novak in the Department of Information Sciences at the University of Texas has been doing research on artificial intelligence that is of particular interest to teachers of physics at the university or secondary school level. He has "taught" his computer to read the problems at the ends of textbook chapters, to construct graphic representations of the problems, when appropriate, and to solve the problems. Out of this work he finds that if the computer constructs a good graphic representation, it is more likely to solve the problem. There may be some value in returning to the practice of asking students to make sketches or diagrams that depict their understanding of a problem.

Computer technology will be brought to bear on the task of improving students' ability to learn science and to more rapidly acquire some level of scientific sophistication. In addition, it is yielding information that may help to improve instruction in the ordinary classroom. (See the discussion in Chapter 7 regarding use of the computer to enhance laboratory work, primarily through simulation.)

POSSIBILITIES

Simulation Experiences

As the sizes of computer memories and computational power have increased, computer programs have become more and more sophisticated. And as these programs become more sophisticated, you can make more uses of the computer because it comes closer to using language like yours. Computer programs can now be used to adjust your car's engine, to create the sophisticated animations you see in ads on your TV screen and in the *Star Wars* films, to synthesize speech ("Speak and Spell"), to "understand" speech to a limited degree, to play a surprisingly good game of chess, and to simulate a wide variety of complex processes. A computer can even simulate how a computer works! A computer simulation can put you, the student, or the entire class in the driver's seat. For example, we can simulate the effect on the economy and on the public health of various environmental regulations and reliance upon various sources of energy. The class can be asked after debate to choose a specific course of action—e.g., switch to coal for energy and eliminate air quality standards for smokestack emissions. Then we can simulate the results of that decision according to the model of the energy and health system that we have built into the computer program. The simulation might show you how murky the sky would look and how many people would die from air pollutants *and* how many extra jobs would be created in the economy (if any) as a result of a switch to coal.

A simulation is a dynamic display based on a model. A dynamic display shows changes that occur as a function of time. A model is a *simplified* version of some system or process that helps us to think about how the system works. For example, a Linc trainer simulates the flight of an aircraft by using a model, a simplified version, of the real world of flight.

A simulation can be a powerful teaching tool. In a sense, each of us is constantly building, in our minds, a simulation of the world. Our muscles can unconsciously and precisely enable us to walk up stairs without much thought because we have stored, in our brain and the rest of our nervous system, a model of a stairway. A centerfielder can start at the crack of a bat and run to the spot to catch a fly ball because he has an incredible intuitive simulation of a baseball's trajectory. In a way, we can think of the entire process of education as trying to improve and refine our model of the world, our simulation of the universe.

A student who is operating a computer simulation has a sense of a realistic activity taking place in real time. A computer can be designed to constantly increase the level of skill required in order to function well. All in all, a good computer simulation has many of the properties that make an activity interesting, challenging, and instructive. For example, a simulation of a space war can be exciting and also can teach the relationship between acceleration and fuel consumption, as well as a variety of strategies for dealing with changes in a system. A simulation of a political negotiation can teach the need and value of compromise and the need for allies by showing the consequences of different strategies. A simulation of a chemical manufacturing plant can teach about many facets of chemical reactions. A simulation is particularly powerful for showing complex relationships in systems that are in transition from one stage to another. As the student acquires skill in a well-designed simulation, the student is also improving the simulation in her or his mind.

As the class becomes more sophisticated, someone is sure to ask, "How do we know that the simulation is correct? How do we know that the model is right? Maybe some medical discovery will make it okay to breathe air with sulfur in it. Maybe we will discover some cheap way to turn coal into gasoline." Such a challenge actually illustrates a strength of computer simulation. *Not only do simulations allow us to see how the system works under a given set of assumptions; they also allow us to modify the assumptions and laws, and see what happens as a result.* A current simulation that is being widely used in colleges and is being modified for precollege use involves population growth. With this simulation we can explore the results of various levels of fertility and discuss the impact that changing population–age patterns will have upon a country. We can project the time when, under current rates of growth, Mexico will have a population equal to that of the United States. We can also explore how much we have to change family-size patterns in order to modify the picture.

In physics, we can create and animate a world in which there is no friction and see what would happen when we try to walk, drive, get out of bed, fix the roof, etc. On a more sophisticated level, we can see what would happen to the solar system if we modified the Law of Universal Gravitation or if the temperature of the earth dropped just 2 degrees Celsius. The student can be asked to carry out some task or operation in these newly simulated worlds on the basis of her or his understanding of how the world works. When expectations are either fulfilled or violated, differences of judgment can be the basis of lessons that refine the students' knowledge and intuition. Much of education enables us to answer the "What if . . .?" question for a variety of situations. Through computers we can compare the answers we get in our minds with the results of a model that someone has built into a computer.

In every phase of the curriculum, we can strive to create interactive experiences that are challenging and engrossing. For example, a recent edition of *Purser's Magazine*[6] has a review of a program entitled "Three Mile Island" that starts, "I thought this was a game. It is not." Muse's "Three Mile Island" is an educational program. It is a simulation of a nuclear power plant. And it is very well done.

Context and Questions

What will the availability of simulations mean to science education? Will the entry of computers mean the end of the classroom or the laboratory? How will the simulation experience affect the mental development of students?

In our judgment, simulations will enrich learning. However, simulation experiences may change the relative emphasis on and importance of particular skills and knowledge. For example, there will be greater emphasis on problem solving and less emphasis on memorization of verbal chains and computational algorithms. There will be more emphasis on spontaneous and precise use of language and symbols because of the need to give very precise commands to the computer.

Real-life laboratory work will still be important because, no matter how complex, a simulation is not the same as reality. A simulation of a chemistry experiment does not have the smell of a chemistry lab; it does not have the same impact as when something goes wrong and acid starts to eat its way into some material on the lab table. It will continue to be important to have our students wrestle with complex, messy reality if they are to appreciate the power and limitations of science and simulations.

Group discussions led by the teacher will continue to be important for the following reasons:

1. As an adjunct to the discussion of complex ideas, the computer provides a basis for exploring more complex phenomena than would otherwise be possible. However, students need to penetrate to the assumptions and values that underly each simulation before they accept the results.

2. Conversation has a richness that, so far, dialogue with a computer lacks. Discussion serves many intellectual and instructional purposes that cannot be served by a computer alone. The possible decision routes one can take in a simulation provide one basis for discussion and arguments that go to the very heart of scientific inquiry.

3. The computer can be used as a powerful medium for exploring dynamic systems (e.g., ecological systems) by the teacher and the class. The class can discuss what decisions should be made in a simulation that controls a simulation. Graphs and animations that would be tedious or impossible to bring into the classroom can be brought up on a large screen and modified as called for as a result of questions and discussions.

From the last two examples we can see that we needn't think in terms of teacher *vs.* computers but rather as teachers *and* computers.

An important aspect of computer simulation work is that the student must constantly be creating ideas in order to drive the simulation. She or he must construct sentences, make decisions, and solve all kinds of problems. Thus, a computer simulation constantly exercises the higher-order intellectual skills that are often neglected because of the limitations of the print medium. An important option is to have the student make up her or his own simulation.

At the present time, detailed images or pictures take up so much memory that we cannot include them as part of a computer program without some help. However, there is a relatively new and inexpensive source of pictorial images—the videodisc—that promises to change this. A single videodisc stores up to 54,000 individual pictures on a single side of a piece of plastic that resembles a silvery phonograph record. Each picture or "frame" can be called up by the computer. Thus, the combination of a computer and a videodisc permits simulation of things that require realistic pictures. For example, we can simulate a trip through an automobile engine and watch the operation of various parts including the explosions in the cylinders. We can simulate a trip to a Mexican village in which we can talk to the villagers almost as if we were in a real village. This kind of vicarious travel can be given the dimensions of an adventure in which certain decisions lead to difficult and dangerous situations.

How would large amounts of time spent on computer simulations affect the minds of students? We think this is an important question. It seems to us that one of the highest priorities should be to try to understand the impact of large amounts computer experience on the psychological and intellectual development of students. We see that computer simulations have the potential to add a significant new dimension, a challenging realism, to instruction. Are there some side effects that we should be wary of? Are there some particular benefits from certain activities that we should capitalize upon in science instruction? (See, for example, the studies described by Hegarty in Chapter 7.)

Given that appropriate software can be developed to fully exploit the newest computer technologies, there is now a potential to go to the heart of science to inquire about what is important in ways that earlier technologies have not readily permitted. Participation in simulations and in programming requires an active stance by the student rather than a passive reception learning set. The challenge presented by the availability of this evolving technology is enormous—and is yet to be fully understood. For people who like adventure, this is the area for exploration in the decade of the 1980's.

Chapter 2

1. National Assessment of Educational Progress. *Three National Assessments of Science: Changes in Achievement, 1969–1977.* Denver: Center for Assessment of Educational Progress, 1978.

2. National Assessment of Educational Progress. *Attitudes Toward Science: Selected Results from the Third National Assessment of Science.* Denver: NAEP, 1979.

3. Deutsch, M. "Cooperation and Trust: Some Theoretical Notes." *Nebraska Symposium on Motivation.* (Edited by M. Jones.) Lincoln: University of Nebraska Press, 1962. pp. 275–320.

Johnson, D.W., and Johnson, R. *Learning Together and Alone: Cooperation, Competition, and Individualization.* Englewood Cliffs, N.J.: Prentice-Hall, 1975.

4. Johnson, D.W., and Johnson, R. *op. cit.*

5. Johnson, D.W., and Johnson, R. "Student Perceptions of and Preferences for Cooperative and Competitive Learning Experiences." *Perceptual and Motor Skills* 42: 989–990; 1976.

Johnson, R.T. "The Relationship Between Cooperation and Inquiry in Science Classrooms." *Journal of Research in Science Teaching* 13: 55–63; 1976.

6. Nelson, L., and Kagan, S. "Competition: The Star-Spangled Scramble." *Psychology Today* 6: 53–56, 90–91; September 1972.

7. Johnson, R.T. *op. cit.*

Johnson, R.T., and Johnson, D.W. "Type of Task and Student Achievement and Aptitude in Interpersonal Cooperation, Competition, and Individualization." *Journal of Social Psychology* 108: 37–48; 1979.

8. Johnson, D.W., and Johnson, R. "Cooperative, Competitive, and Individualistic Interdependence in the Classroom." *Journal of Research and Development in Education* (Special Issue) 12(1), Fall 1978.

Johnson, D.W., and Johnson, R. *Learning Together and Alone.*

Sharan, S. "Cooperative Learning in Teams: Recent Methods and Effects on Achievement, Attitudes, and Ethnic Relations. *Review of Educational Research* 50: 241–272; 1980.

9. National Assessment of Educational Progress. *Three National Assessments of Science.*

10. Johnson, D.W. *Educational Psychology.* Englewood Cliffs, N.J.: Prentice-Hall, 1979.

11. Johnson, D.W., and others. "The Effects of Cooperative, Competitive, and Individualistic Goal Structures on Achievement: A Meta-analysis." *Psychological Bulletin* (in press).

12. Humphreys, B.; Johnson, R.; and Johnson, D.W. "Cooperation, Competition, Individualization, and the Ninth Grade Science Student." Minneapolis: University of Minnesota, 1981. (Mimeo.)

Johnson, D.W.; Johnson, R.; and Skon, L. "Student Achievement on Different Types of Tasks Under Cooperative, Competitive, and Individualistic Conditions." *Contemporary Educational Psychology* 4: 99–106; 1979.

13. Johnson, D.W., and Johnson, R. "Conflict in the Classroom: Controversy and Learning." *Review of Educational Research* 49: 51–70; 1979.

14. Stake R., and Easley, J. "Case Studies in Science Education." Center for Instructional Research and Curriculum Evaluation and Committee on Culture and Cognition, University of Illinois, 1978.

15. Johnson, D.W., and Johnson, R. "Conflict in the Classroom."

16. Johnson, D.W., and Johnson, R. "Conflict in the Classroom."

Johnson, D.W. "Group Processes: Influence of Student–Student Interaction on School Outcomes." *The Social Psychology of School Learning.* New York: Academic Press, 1980.

17. National Assessment of Educational Progress. *Attitudes Toward Science.*

18. *Minnesota School Affect Assessment Manual.* Minneapolis: Minnesota High School Testing Program, University of Minnesota, 1974.

19. Johnson, D.W., and Johnson, R. "Cooperative, Competitive, and Individualistic Interdependence in the Classroom."

Johnson, D.W., and Johnson, R. *Learning Together and Alone.*

20. Johnson, D.W., and Ahlgren, A. "Relationship Between Student Attitudes About Cooperation and Competition and Attitudes Toward Schooling." *Journal of Educational Psychology* 68: 92–102; 1976.

Johnson, D.W. and others. "The Effects of Cooperative vs. Individualized Instruction on Student Prosocial Behavior, Attitudes Toward Learning, and Achievement." *Journal of Educational Psychology* 68: 446–452; 1976.

21. Johnson, D.W., and Johnson, R. "Cooperative, Competitive, and Individualistic Interdependence in the Classroom."

22. National Assessment of Educational Progress. *Three National Assessments of Science.*

23. National Assessment of Educational Progress. *Attitudes Toward Science.*

24. Johnson, D.W., and Johnson, R. "Cooperative, Competitive, and Individualistic Interdependence in the Classroom."

Johnson, D.W., and Johnson, R. *Learning Together and Alone.*

Johnson, R.T., and Johnson, D.W. "The Social Integration of Handicapped Students into the Mainstream." *Social Environment of the Schools.* (Edited by M. Reynolds.) Reston, Va.: Council for Exceptional Children, 1980. pp. 9–39.

25. Johnson, R.T., and Johnson, D.W. *op. cit.*

26. Remmers and Radler. "Purdue Opinion Panel." *The American Teenager.* Indianapolis–New York: Bobbs Merrill Co., 1957. p. 164.

27. Gunderson, and Johnson, D.W. "Building Positive Attitudes by Using Cooperative Learning Groups." *Foreign Language Annals* 13: 39–46; 1980.

Johnson, D.W., and Ahlgren, A. *op. cit.*

Johnson, D.W., and others. *op. cit.*

Johnson, D.W., and Norem-Hebeisen, A. "Attitudes Toward Interdependence Among Persons and Psychological Health." *Psychological Reports* 40: 843–850; 1977.

Norem-Hebeisen, A., and Johnson, D.W. "Relationship Between Cooperative, Competitive and Individualistic Attitudes and Differentiated Aspects of Self Esteem." *Journal of Personality* (in press).

28. Johnson, D.W.; Johnson, R.; and Scott, L. "The Effects of Cooperative and Individualized Instruction on Student Attitudes and Achievement." *Journal of Social Psychology* 104: 207–216; 1978.

29. Johnson, D.W., and Norem-Hebeisen, A. *op. cit.*

30. Johnson, D.W., and Johnson, R. *Learning Together and Alone.*

31. *Ibid.*

32. *Ibid.*

Chapter 3

1. See, for example: Ryans, D. *Characteristics of Teachers.* Washington, D.C.: American Council on Education, 1960.

2. Stolurow, L. "Model the Master Teacher or Master the Teaching Model." *Learning and the Educational Process.* (Edited by J. Krumboltz.) Chicago: Rand McNally, 1965.

3. Rosenshine, B., and Furst, N. "Research on Teacher Performance Criteria." *Research in Teacher Education.* (Edited by B. Smith.) Englewood Cliffs, N.J.: Prentice-Hall, 1971.

4. Medley, D.; Soar, R.; and Soar, R. *Assessment and Research in Teacher Education: Focus on PBTE.* Washington, D.C.: American Association of Colleges for Teacher Education, 1975.

5. *Ibid.*

6. Simon, A., and Bayer, E. *Mirrors for Behavior III.* Philadelphia: Research for Better Schools, 1974.

7. Borich, G., and Madden, S. *Elementary Classroom Instruction.* Reading, Mass.: Addison-Wesley, 1977.

8. Okey, J., and Capie, W. "Assessing the Competence of Science Teachers." *Science Education* 64: 279–287; 1980.

9. Anderson, R.; James H.; and Struthers, J. "The Teaching Strategies Observation Differential." *Human Interaction in Education.* (Edited by G. Standard and A. Roark.) Boston: Allyn and Bacon, 1974.

10. Yeany, R., and Capie, W. "Analysis System for Describing and Measuring Strategies of Teaching Data Manipulation and Interpretation." *Science Education* 63: 355–361; 1979.

11. Capie, W.; Dillashaw, F.; and Okey, J. "Keeping Students On-Task." *Science Teacher,* December 1979. pp. 31–32.

12. Fox, R.; Luszki, M.; and Schmuck, R. *Diagnosing Classroom Learning Environments.* Chicago: Science Research Associates, 1966.

13. Gage, N. *Teacher Effectiveness and Teacher Education.* Palo Alto, Calif.: Pacific Books, 1972.

14. Sorenson, G., and Husek, T. "Development of a Measure of Teacher Role Expectations." *American Psychologist* 18: 389; 1963.

15. Hord, G. *A Strategy for Evaluation of Teacher Classroom Behavior.* Master's thesis. Austin: The University of Texas, 1971. (Typewritten.)

16. Tollefson, N. "Selected Student Variables and Perceived Teacher Effectiveness." *Education* 94: 30–35; 1973.

17. Rosenshine, B., and Furst, N. "The Use of Direct Observation To Study Teaching." *Second Handbook of Research on Teaching.* (Edited by N. Travers.) Chicago: Rand McNally and Company, 1973.

18. Hart, E., and Towes, W. "A Methodology for Examination of the Relationship Between Student and Teacher Understanding of BSCS Concepts and the BSCS Conceptual Framework." *Journal of Research in Science Teaching* 17 (3): 251–256; 1980.

19. Lamb, W., and others. "The Effect on Student Achievement of Increasing Kinetic Structure of Teachers' Lectures." *Journal of Research in Science Teaching* 16 (3): 223–228; 1979.

20. Butts, D., and Hord, S. *Personalizing Classroom Interaction: A Self Directed Strategy for Cooperating Teachers and Advising Teachers.* Austin: The University of Texas, The Research and Development Center for Teacher Education, 1972.

21. Ashley, J., and Butts, D. "A Study of the Impact of an Inservice Education Program on Teaching Behavior." *Research and Curriculum Development in Science Education.* (Edited by D. Butts.) Austin: The University of Texas, 1970.

Flanders, N. *Teacher Influence, Pupil Attitudes, and Achievement.* U.S. Department of Health, Education, and Welfare, Office of Education, Cooperative Research Monograph No. 12. Washington, D.C.: Government Printing Office, n.d.

22. Shulman, L., and Tamir, P. "Research of Teaching in the Natural Sciences." *Handbook of Research on Teaching.* (Edited by R. Travers.) Chicago: Rand McNally and Company, 1973.

23. Schwab, J. "The Teaching of Science as Inquiry." *The Teaching of Science.* (Edited by J. Schwab and P. Brandwein.) Cambridge, Mass.: Harvard University Press, 1962.

24. Tollefson, N. *op. cit.*

25. Rosenshine, B., and Furst, N. "Research on Teacher Performance Criteria."

26. Piper, M. "A Science Activity Teaching Plan." *School Science and Mathematics* 80 (5): 399–406; 1980.

27. Kounin, J.S. *Discipline and Group Management in the Classroom.* New York: Holt, Rinehart and Winston, 1970.

28. Campbell, J. "Science Teachers' Flexibility." *Journal of Research in Science Teaching* 14 (6): 525–532; 1977.

29. Medley, D. "The Effectiveness of Teachers." *Research in Teaching.* (Edited by R. Peterson and H. Walberg.) Berkeley, Calif.: McCutchan Publishing Corp., 1979.

30. Shymansky, J., and others. "A Study of Student Classroom Behavior and Self-Perception as It Relates to Problem Solving." *Journal of Research in Science Teaching* 14 (3): 191–198; 1977.

31. Medley, D. *op. cit.*

32. Rosenshine, B., and Furst, N. "The Use of Direct Observation To Study Teaching."

33. Capie, W.; Dillashaw, F.; and Okey, J. *op. cit.*

34. Boulanger, F. "Relationship of an Inservice Program to Student Learning: Naturalistic Documentation." *Science Education* 64 (3): 349–356; 1980.

35. McDuffie, T., and Beehler, C. "Achievement–Workstyle Relationships in ISCS Level I." *Journal of Research in Science Teaching* 15 (6): 485–490; 1978.

36. Penick, J., and Shymansky, J. "The Effects of Teacher Behavior on Student Behavior in Fifth-Grade Science: A Replication Study." *Journal of Research in Science Teaching* 14(5): 427–431; 1977.

37. Test, D., and Heward, W. "Photosynthesis: Teaching a Complex Science Concept to Juvenile Delinquents." *Science Education* 64 (2): 129–140; 1980.

38. Koran, J.; Koran, M.; and Baker, S. "Differential Response to Cueing and Feedback in

the Acquisition of an Inductively Presented Biological Concept." *Journal of Research in Science Teaching* 17 (2): 167–172; 1980.

39. Berliner, D. "Impediments to the Study of Teacher Effectiveness." Paper presented at the Conference on Research on Teacher Effects: An Examination by Decision-Makers and Researchers, The University of Texas, Austin, Texas, November 2–4, 1975.

40. Raven, R., and Cole, R. "Relationships Between Piaget's Operative Comprehension and Physiology Modeling Processes of Community College Students." *Science Education* 62 (4): 481–490; 1978.

41. Martin, W. "The Use of Behavioral Objectives in Instruction of Basic Vocational Science Students." *Journal of Research in Science Teaching* 14 (1): 1–12; 1977.

42. Butts, D., and Hord, S. *op. cit.*

43. Gage, N. *op. cit.*

44. Sorenson, G., and Husek, T. *op. cit.*

45. Campbell, J. "Can a Teacher Really Make a Difference?" *School Science and Mathematics* 74: 657–666; 1974.

46. Medley, D. *op. cit.*

47. Rosenshine, B. "Recent Research on Teaching." Paper presented at the Conference on Research on Teacher Effects: An Examination by Decision-Makers and Researchers. The University of Texas, Austin, Texas, November 2–4, 1975.

48. Gabel, D., and Herron, J. "The Effects of Grouping and Pacing on Learning Rate, Attitude and Retention in ISCS Classrooms." *Journal of Research in Science Teaching* 14 (5): 385–400; 1977.

49. Rice, M., and Linn, M. "Study of Student Behavior in a Free Choice Environment." *Science Education* 62 (3): 365–376; 1978.

50. Linn, M.; Chen, B.; and Their, H. "Teaching Children To Control Variables: Investigation of a Free-Choice Environment." *Journal of Research in Science Teaching* 14 (3): 249–256; 1977.

51. Rosenshine, B. *op. cit.*

52. Butts, D., and Hord, S. *op. cit.*

53. Bryant, B. "Locus of Control Related to Teacher–Child Interperceptual Experiences." *Child Development* 45: 157–164; 1974.

Murray, H.; Herling, G.; and Staebler, B. "The Effects of Locus of Control and Pattern of Performance on Teachers' Evaluation of a Student." *Psychology in the Schools* 10: 345–350; 1973.

54. Campbell, J. "Can a Teacher Really Make a Difference?"

55. Rosenshine, B., and Furst, N. "Research on Teacher Performance Criteria."

56. Butts, D., and Hord, S. *op. cit.*

57. Brown, J., and MacDougall, M. "Teacher Consultation for Improved Feeling of Self-Adequacy in Children." *Psychology in the Schools* 10: 320–326; 1973.

58. Campbell, J. *op. cit.*

59. Sorenson, G., and Husek, T. *op. cit.*

60. Gage, N. *op. cit.*

61. Tollefson, N. *op. cit.*

62. Rosenshine, B., and Furst, N. "The Use of Direct Observation To Study Teaching."

63. Edwards, C. "Relation of Productivity in Elementary School Science to Token Reinforcement Involving Liked and Unliked Peers." *Journal of Research in Science Teaching* 14 (5): 449–454; 1977.

64. Edwards, C., and Surma, M. "The Relationship Between Type of Teacher Reinforcement and Student Inquiry Behavior in Science." *Journal of Research in Science Teaching* 17 (4): 337–342; 1980.

65. Rosenshine, B. *op. cit.*

66. Butts, D., and Hord, S. *op. cit.*

67. Alpert, J. "Teacher Behavior Across Ability Groups: A Consideration of the Mediation of Pygmalion Effects." *Journal of Educational Psychology* 66: 348–353; 1974.

Ashley, J., and Butts, D. *op. cit.*

Campbell, J. *op. cit.*

Flanders, N. *op. cit.*

Gage, N. *op. cit.*

Good, T., and Brophy, J. "Changing Teacher and Student Behavior: An Empirical Investigation." *Journal of Educational Psychology* 66: 390–405; 1974.

Samph, T. "Teacher Behavior and the Reading Performance of Below-Average Achievers." *Journal of Educational Research* 67: 268–270; 1974.

Tollefson, N. *op. cit.*

68. Mueller, D. "The Second-Round Question (or How Teachers React to Student Responses)." *Urban Education* 21: 153–165; 1973.

69. Arnold, D.; Atwood, R.; and Rogers, V. "An Investigation of Relationships Among Question Level, Response Level and Lapse Time." *School Science and Mathematics* 73: 591–594; 1973.

70. Rosenshine, B., and Furst, N. *op. cit.*

71. Medley, D. *op. cit.*

72. Rosenshine, B., and Furst, N. "Research on Teacher Performance Criteria."

73. DeTure, L. "Relative Effects of Modeling on the Acquisition of Wait-Time by Preservice Elementary Teachers and Concomitant Changes in Dialogue Patterns." *Journal of Research in Science Teaching* 16 (6): 553–562; 1979.

74. Rowe, M.B. "Relation of Wait-Time and Rewards to Development of Language, Logic, and Fate Control: Part One—Wait-Time." *Journal of Research in Science Teaching* 11 (2): 81–89; 1974.

75. Rice, D. "The Effect of Question-Asking Instruction on Preservice Elementary Science Teachers." *Journal of Research in Science Teaching* 14 (4): 353–360; 1977.

76. Tobin, K. "The Effect of an Extended Teacher Wait-Time on Science Achievement." *Journal of Research in Science Teaching* 17 (5): 469–476; 1980.

77. Rosenshine, B. *op. cit.*

78. DeBoer, G. "Can Repeated Testing of *En Route* Objectives Improve End-of-Course Achievement in High School Chemistry?" *Science Education* 64 (2): 141–148; 1980.

79. Burrows, C., and Okey, J. "The Effects of a Mastery Learning Strategy on Achievement." *Journal of Research in Science Teaching* 16 (1): 33–38; 1979.

80. Yeany, R., and Capie, W. *op. cit.*

81. Pouler, C., and Wright, E. "An Analysis of the Influence of Reinforcement and Knowledge of Criteria on the Ability of Students To Generate Hypotheses." *Journal of Research in Science Teaching* 17 (1): 31–38; 1980.

82. Medley, D. *op. cit.*

83. Rosenshine, B. *op. cit.*

84. Riley, J. "Effects of Studying a Question Classification System on the Cognitive Level of Preservice Teachers' Questions." *Science Education* 62 (3): 333–338; 1978.

Chapter 4

1. Louis Harris Poll. "Family Weekly." *Gainesville Sun*, April 1980.

2. "The Power of Informal Science Education." *Outdoor Biology Instructional Stategies (OBIS) Newsletter.* Lawrence Hall of Science, March 1980.

3. Templeton, Michael. "Industry Sponsored Exhibits Are Useful Educational Tools." *Science and Children* 17 (6): 11–12; March 1980.

4. Laetch, Watson M., and others. "Children and Family Groups in Science Centers." *Science and Children* 17 (6): 14–17; March 1980.

5. Shettel, H.H. "An Evaluation of Visitor Response to Man in His Environment—Final Report." Chicago: Field Museum of Natural History; Washington, D.C.: American Institute for Research in the Behavioral Sciences, July 1968.

6. Tobias, S. "Achievement-Treatment Interaction." *Review of Educational Research* 46: 61–74; 1975.

7. Koran, Mary, and Koran, John J., Jr. "The Study of Aptitude Treatment Interaction: Implications for Educational Communications and Technology." *Instructional Communications and Technology Research Reports.* (Edited by Francis M. Dwyer.) Vol. 9, No. 3, January 1979.

8. Cronbach, L.J., and Snow, R.E. *Aptitudes and Instructional Methods.* New York: Irvington, 1977.

9. Keele, Stephen W. *Attention and Human Performance.* Pacific Palisades, Calif.: Goodyear Publishing Company, Inc., 1973.

10. Screven, C.G. *The Measurement and Facilitation of Learning in the Museum Environment: An Experimental Analysis.* Washington, D.C.: Smithsonian Institution Press, 1974.

11. DeWoard, Richard J., and others. "Effects of Using Programmed Cards on Learning in a Museum Environment." *Journal of Educational Research* 67 (10): 457–460; 1974.

12. Novak, Joseph D. "Understanding the Learning Process and Effectiveness of Teaching Methods in Classroom, Laboratory, and Field." *Science Education* 60 (4): 493–512; 1979.

13. Anderson, R.C. "Control of Student Mediating Processes During Verbal Learning and Instruction." *Review of Educational Research* 40 (3): 349–371; 1970.

Rothkopf, E.Z. "The Concept of Mathemagenic Activities." *Review of Educational Research* 40 (3): 325–336; 1970.

14. Richards, J.P. "Adjunct Post Questions in Text: A Critical Review of Methods and Processes." *Review of Educational Research* 49: 181–196; 1979.

15. Koran, John J., Jr., and others. "Evaluating a Walk Through Science Exhibit: The Florida Cave." Manuscript in preparation, Florida State Museum and College of Education, University of Florida, Gainesville, 1980.

16. Niehoff, A. "Characteristics of the Audience Reaction in the Milwaukee Public Museum." *The Museum Visitor.* (Edited by S.F. de Berhegyi and I.A. Hanson.) Publications in Museology 3. Milwaukee: Milwaukee Public Museum, 1968.

Niehoff, A. "Audience Reaction in the Milwaukee Public Museum: The Winter Visitor." *The Museum Visitor.* (Edited by S.F. de Berhegyi and I.A. Hanson.) Publications in Museology 3. Milwaukee: Milwaukee Public Museum, 1968.

17. Cohen, Marilyn S., and others. "Orientation in a Museum—An Experimental Visitors' Survey." *Curator* 20 (2): 85–97; June 1977.

18. Washburne, R. F., and Wagar, J. A. "Evaluating Visitor Response to Exhibit Content." *Curator* 15 (3): 248–254; 1972.

19. Melton, Arthur. "Visitor Behavior in Museums: Some Early Research in Environmental Design." *Human Factors* 14 (4): 393–403; 1972.

20. Nielson, L. C. "A Technique for Studying the Behavior of Museum Visitors." *Journal of Educational Psychology* 37: 103–110; 1946.

21. Wittlin, Alma. "Exhibits: Interpretive, Underinterpretive, Misinterpretive—Absolutes in Exhibit Techniques." *Museums and Education.* (Edited by Eric Larrabee.) Washington, D.C.: Smithsonian Press, 1968.

22. Oppenheimer, F. "The Exploratorium: A Playful Museum Combines Perception and Arts and Science Education." *American Journal of Physics* 40 (7): 978–984; 1972.

Oppenheimer, F., and Cole, K.C. "The Exploratorium: A Participatory Museum." *Prospectus* 4 (1): 4–10; 1974.

23. Hannibal, Emmett. "The Wadsworth Atheneum Tactile Gallery." *The Art Museum as Educator.* (Edited by Barbara Newsome.) Berkeley: University of California Press, 1978.

Hoth, Sue Robinson. "The National Museum of Natural History: The Discovery Room." *The Art Museum as Educator.* (Edited by Barbara Newsome.) Berkeley: University of California Press, 1978.

24. Eason, Laurie P., and Linn, Marcia C. "Evaluation of the Effectiveness of Participatory Exhibits." *Curator* 19 (1): 45–62; March 1976.

25. Arth, Malcolm, and Claremon, Linda. "The Discovery Room." *Curator* 20 (3): 169–180; September 1977.

26. Tard, G. *The Laws of Imitation.* New York: Holt, Rinehart, and Winston, 1903.

27. Bandura, Albert, and Walters, R. *Social Learning and Personality Development.* Chicago: Holt, Rinehart and Winston, 1963.

28. Koran, John J., Jr. "The Use of Modeling, Feedback, and Practice Variables to Influence Teacher Behavior." *Science Education* 56 (3): 285–291; 1972.

29. Koran, John J., Jr., and others. "Studying Modeling in Science Museums: Two Studies." Manuscript in preparation, Florida State Museum and College of Education, University of Florida, Gainesville, 1980.

30. Shettel, H.H. *op. cit.*

31. Tobias, S. *op. cit.*

32. Cronbach, L.J., and Snow, R.E. *op. cit.*

33. Koran, Mary, and Koran, John J., Jr. *op. cit.*

34. Screven, C.G. *op. cit.*

35. DeWoard, Richard J., and others. *op. cit.*

36. Koran, John J., Jr., and Baker, S. Dennis. "Evaluating Effectiveness of Field Experiences." *What Research Says to the Science Teacher.* (Edited by M. B. Rowe.) Washington, D.C.: National Science Teachers Association, 1979. Volume 2.

Chapter 5

1. Johnson, Roger T.; Ryan, Frank L.; and Schroeder, Helen. "Inquiry and the Development of Positive Attitudes." *Science Education* 58 (1): 51–56; January–March 1974.

2. Metz, William Charles. "The Effects of Two Modes of Instruction on the Curiosity and the Attitudes Toward Science of Elementary School Children." *Dissertation Abstracts* 37 (1): 123A; 1976.

3. Davis, Maynard. "The Effectiveness of a Guided-Inquiry Discovery Approach in an Elementary School Science Curriculum." *Dissertation Abstracts*, 39 (4): 4164A; 1978.

4. Peterson, Penelope L. "Direct and Open Instructional Approaches: Effective for What and For Whom?" Working Paper No. 243. Wisconsin Research and Development Center for Individualized Schooling, October 1978.

5. *Ibid.*

6. Today most activity-based kit programs include science textbooks for students.

7. Wellman, Ruth. *What Research Says to the Science Teacher.* (Edited by M. B. Rowe.) Washington, D.C.: National Science Teachers Association, 1978. Volume 1.

Additional Reading

Benson, Keith Sheran. "A Comparison of Two Methods of Teaching First Grade Science." *Dissertation Abstracts* 30 (3): 1067A; 1968.

Billings, Gilbert Wendell. "The Effects of Verbal Introduction of Science Concepts on the Acquisition of These Concepts by Children at the Second Grade Level." *Dissertation Abstracts* 36 (10): 6578A; 1976.

Blomberg, Karin Josefina. "A Study of the Effectiveness of Three Methods for Teaching Science in the Sixth Grade." *Dissertation Abstracts* 35 (5): 3290A; 1974.

Byrne, Katherine Ann. "A Cross-Age Study in Elementary School Science." *Dissertation Abstracts* 33 (4): 1536A; 1972.

Fuller, Ellen White. "The Science Achievement of Third Graders Using Visual, Symbolic, and Manipulative Instructional Treatments." *Dissertation Abstracts* 38 (6): 6633A; 1977.

Kemp, Judy Beth. "An Investigation of the Effects of Varying Left and Right Hemisphere Activities on the Achievement of Fifth-Grade Science Students." *Dissertation Abstracts* 39 (5): 5430A; 1978.

MacBeth, Douglas Russell. "The Extent to Which Pupils Manipulate Materials and Attainment of Process Skills in Elementary School Science." *Journal of Research in Science Teaching* 11 (1): 45-51; 1974.

Marlins, James Gregory. "A Study of the Effects of Using the Counterintuitive Event in Science Teaching on Subject Matter Achievement and Subject Matter Retention of Upper-Elementary School Students." *Dissertation Abstracts* 34 (5): 2413A; 1973.

Voelker, Alan M. "Elementary School Children's Attainment of the Concepts of Physical and Chemical Change — A Replication." *Journal of Research in Science Teaching* 12 (1): 5-14; January 1975.

Vongchusiri, Pricha. "The Effects of Alternative Instructional Methods on the Achievement of Science Students in the Elementary Schools of Northeast Thailand." *Dissertation Abstracts* 35 (7): 4318A; 1974.

Chapter 6

1. National Assessment of Educational Progress. "Attitudes Toward Science: A Summary of Results from the 1976-77 National Assessment of Science." Report No. 08-S-02, 1979.

2. Barber, B. *Science and the Social Order.* New York: Collier Books, 1962.

Haney, Richard E. "The Development of Scientific Attitudes." *Science Teacher* 31: 33-35; 1964.

Koslow, M. James, and Nay, Marshall A. "Measuring Scientific Attitudes." *Science Education* 60 (2): 147-172; 1976.

3. Peterson, Rita W. "Changes in Curiosity Behavior from Childhood to Adolescence." *Journal of Research in Science Teaching* 16 (3): 185-192; 1979.

4. Peterson, Rita W. "The Differential Effect of an Adult's Presence on the Curiosity Behavior of Children." *Journal of Research in Science Education* 12 (3): 199-208; 1975.

5. National Assessment of Educational Progress. *op. cit.*

6. Billeh, Victor, and Zachariades, George A. "The Development and Application of a Scale of Measuring Scientific Attitudes." *Science Education* 59 (2): 155-165; 1975.

7. Blatt, Marvin H. "An Investigation of Two Methods of Science Instruction and Teacher

Attitudes Towards Science." *Journal of Research in Science Teaching* 14 (6): 533–538; 1977.
    8. *Ibid.*
    9. Shrigley, Robert. "The Function of Professional Reinforcement in Supporting a More Positive Attitude of Elementary Teachers Toward Science." *Journal of Research in Science Teaching* 14 (4): 317–322; 1977.
    10. Berger, Carl F. "Investigation of Teacher Behavior: Interaction with New Curriculum Material." *Attitudes Toward Science: Investigations.* Columbus, Ohio: SMEAC, 1977. pp. 64–74.
    11. Berger, Carl F. *op. cit.*
    Piper, Martha K., and Hough, Linda. "Attitudes and Openmindedness of Undergraduate Students Enrolled in a Science Methods Course and a Freshman Physics Course." *Journal of Research in Science Teaching* 16 (3): 193–197; 1979.
    12. Piper, Martha K., and Hough, Linda. *op. cit.*
    13. Blatt, Marvin H. *op. cit.*
    14. Earl, Robert D., and Winklejohn, Dorothy R. "Attitudes of Elementary Teachers Toward Science and Science Teaching." *Science Education* 61 (1): 41–45; 1977.
    15. Berger, Carl F. *op. cit.*
    16. Lazarowitz, Reuven. "Does Use of Curriculum Change Teachers' Attitudes Towards Inquiry?" *Journal of Research in Science Teaching* 13 (6): 547–552; 1976.
    17. Rowe, Mary B. "Relation of Wait-Time and Rewards to the Development of Language, Logic and Fate Control: Part II—Rewards." *Journal of Research in Science Teaching* 2 (2): 291–308; 1974.
    Rowe, Mary B. *Teaching Science as Continuous Inquiry.* New York: McGraw-Hill Book Co., 1973.
    18. Rowe, Mary B. *Teaching Science as Continuous Inquiry.*
    19. Johnson, Roger; Ryan, Frank; and Schroeder, Helen. "Inquiry and the Development of Positive Attitudes." *Science Education* 58 (1): 51–56; 1974.
    20. Krockover, Gerold, and Malcolm, Marshall. "The Effects of the Science Curriculum Improvement Study on a Child's Self-Concept." *Journal of Research in Science Teaching* 14 (4): 295–299; 1977.
    21. Lawrenz, Frances. "The Prediction of Student Attitude Toward Science from Student Perception of the Classroom Learning Environment." *Journal of Research in Science Teaching* 13 (6): 509–515; 1976.
    22. McMillian, James H., and May, Marcia J. "A Study of Factors Influencing Attitudes Toward Science of Junior High School Students." *Journal of Research in Science Teaching* 16 (3): 217–222; 1974.
    23. Peterson, Rita W. *op. cit.*
    24. Sabar, Naama, and Kaplan, Eugene. "The Effect of a New Seventh-Grade Biology Curriculum on the Achievements and Attitudes of Intellectually and Culturally Heterogeneous Classes." *Journal of Research in Science Teaching* 15 (4): 271–276; 1978.

Chapter 7

    1. Champagne, A.B., and Klopfer, L.E. *Cumulative Index to "Science Education": Volumes 1 through 60, 1916–1976.* New York: Wiley, 1978.
    2. Chambers, R.G. "What Use Are Practical Physics Classes?" *Bulletin of the Institute of Physics and Physical Society* 14: 181–183; 1963.
    3. Boud, D.J., and others. *Laboratory Teaching in Tertiary Science: A Review of Some Recent Developments.* Sydney: Higher Education, Research and Development Society of Australia, 1978.
    4. Postelthwait, S.N.; Novak, J.; and Murray, H.T., Jr. *The Audio-Tutorial Approach to Learning: Through Independent Study and Integrated Experiences.* Minneapolis: Burgess, 1969.
    5. Keller, F.S. "Good-Bye, Teacher." *Journal of Applied Behaviour Analysis* 1: 79–89; 1968.
    6. Harnischfeger, A., and Wiley, D.E. "The Teaching–Learning Process in Elementary Schools: A Synoptic View." *Curriculum Inquiry* 6: 5–43; 1976.
    Johnson, M. "The Translation of Curriculum into Instruction." *Journal of Curriculum Studies* 1: 115–131; 1969.
    Saylor, J.G., and Alexander, W.M. *Planning Curriculum for Schools.* New York: Holt, Rinehart and Winston, 1974.

7. Schwab, J.J. "The Practical: A Language for Curriculum." *School Review* 78: 1–23; 1969.
Schwab, J.J. "The Practical 2: Arts of Eclectic." *School Review* 79: 493–542; 1971.
Schwab, J.J. "The Practical 3: Translation into Curriculum." *School Review* 81: 501–522; 1973.
Schwab, J.J. "Decision and Choice: The Coming Duty of Science Teaching." *Journal of Research in Science Teaching* 11: 309–317; 1974.

8. Klopfer, L.E. "Evaluation of Learning in Science." *Handbook of Formative and Summative Evaluation of Student Learning.* (Edited by B.S. Bloom, J.T. Hastings, and G.F. Madaus.) New York: McGraw-Hill, 1971.

9. Fraser, B.J. "Are ASEP's Stated Aims Worth Achieving?" *Australian Science Teachers' Journal* 22 (3): 130–132; 1976.

10. Bates, G.C. *The Role of the Laboratory in Secondary School Science Programs.* (Edited by M.B. Rowe.) What Research Says to the Science Teacher Series. Washington, D.C.: National Science Teachers Association, 1978. Volume 1.
Bradley, R.L. "Is the Science Laboratory Necessary for General Education Science Courses?" *Science Education* 52: 58–66; 1968.
Cunningham, H.A. "Lecture Demonstration Versus Individual Laboratory Method in Science Teaching—A Summary." *Science Education* 30: 70–82; 1946.

11. Kogut, M. "Sticks and Stones . . .!" *Biochemical Education* 4: 9–10; 1976.

12. Ausubel, D.P. *Educational Psychology: A Cognitive View.* New York: Holt, Rinehart and Winston, 1968.
Ausubel, D.P.; Novak, J.D.; and Hanesian, H. *Educational Psychology: A Cognitive View.* Second edition. New York: Holt, Rinehart and Winston, 1978.
Novak, J.D.; Ring, D.G.; and Tamir, P. "Interpretation of Research Findings in Terms of Ausubel's Theory and Implications for Science Education." *Science Education* 55: 483–526; 1971.

13. Keller, F.S. *op. cit.*

14. Postelthwait, S.N.; Novak, J.; and Murray, H.T., Jr. *op. cit.*

15. Case, C.L. "Impact of Audio-Tutorial Laboratory Instruction in Biology on Student Attitudes." *American Biology Teacher* 42: 121–123; 1980.
Case, C.L. "The Influence of Modified Laboratory Instruction on College Student Biology Achievement." *Journal of Research in Science Teaching* 17: 1–6; 1980.
Grobe, C.H., and Sturges, A.W. "The Audio-Tutorial and Conventional Methods of College Level Biology for Non-Science Majors." *Science Education* 57: 499–516; 1973.
Rowsey, R.E., and Mason, W.H. "Immediate Achievement and Retention in Audio-Tutorial Versus Conventional Lecture–Laboratory Instruction." *Journal of Research in Science Teaching* 12: 393–397; 1975.

16. Brown, G.C., and others. "Self-Paced Introductory Physics Course—An Eight-Year Progress Report." *American Journal of Physics* 45: 1082–1088; 1977.
Silberman, R. "The Keller Plan: A Personal View." *Journal of Chemical Education* 55: 97–98; 1978.

17. Doty, R.B. "Personalized Laboratory Instruction in Microbiology: The Audio-Tutorial Approach." *Educational Technology* 14: 36–39; 1974.
Dowdeswell, W.H. "Independent Learning and Biology Teaching in Britain and the U.S.A." *Journal of Biological Education* 7: 8–15; 1973.
Ellias, L.C.; Wildman, T.M.; and Towle, N.J. "Educational Technology and Microbiology." *Journal of College Science Teaching* 6 (1): 30–32; 1976.
Hackett, D., and Holt, I.V. "Biological Science as an Audio-Tutorial System of Instruction for the Non-Science Major." *Science Education* 57: 499–516; 1973.
Von Blum, R. "Individualizing Instruction in Large Undergraduate Biology Laboratories." *American Biology Teacher* 37: 467–469; 1975.
Voyles, M.M., and Wright, E.R. "Individualization in a Large Introductory Microbiology Course." *Science Education* 59: 1–4; 1975.

18. Brewer, I.M. "Recall, Comprehension and Problem Solving." *Journal of Biological Education* 8: 101–112; 1974.
Brewer, I.M. "SIMIG: A Case Study of an Innovative Method of Teaching and Learning." *Studies in Higher Education* 2 (1): 33–54; 1977.

19. Butzow, J.W.; Linz, L.W.; and Drake, R.A. "A Study of the Interrelationships of Attitude and Achievement Measures in an Audio-Tutorial College Chemistry Course." *Journal of Research in Science Teaching* 14: 45–49; 1977.

20. Brown, G.C., and others. *op. cit.*

21. Cassen, T., and Forster, L. "A Self-Paced Laboratory Course for Non-Science Majors." *Journal of Chemical Education* 50: 560–561; 1973.

Peterson, D.L. "Self-Paced Quant: A Keller-Type Offering of Quantitative Analysis or the Keller Plan the Hard Way." *Journal of Chemical Education* 54: 362–364; 1977.

Valeriote, I.M. "A Self-Paced Course in First-Year General Chemistry." *Journal of Chemical Education* 53: 106–108; 1976.

22. Behroozi, F. "Freshman Physics Laboratory in a Small College Overseas." *American Journal of Physics* 44: 334–336; 1976.

Kahn, P.B., and Strassenburg, A.A. "Instructional Innovation in Physics at Stony Brook." *American Journal of Physics* 43: 400–407; 1975.

23. Ott, M.D., and Macklin, D.B. "A Trait–Treatment Interaction in a College Physics Course." *Journal of Research in Science Teaching* 12: 111–119; 1975.

24. Fitts, P.M., and Posner, M.I. *Human Performance.* Belmont, Calif.: Wadsworth, Brooks-Cole, 1967.

25. Hegarty, E.H. "How To Organize Effective Laboratory Teaching in Medicine: Part 1, Purposes." *Medical Teacher* 1: 175–181; 1979.

26. Boud, D.J. "The Laboratory Aims Questionnaire—A New Method for Course Improvement?" *Higher Education* 2: 81–94; 1973.

27. Lynch, P.P., and Gerrans, G.C. "The Aims of First Year Chemistry Courses, the Expectations of New Students and Subsequent Course Influences." *Research in Science Education* 7: 173–180; 1977.

28. Runquist, O. "Programmed Independent Study, Laboratory Technique Course for General Chemistry." *Journal of Chemical Education* 56: 616–617; 1979.

29. Beasley, W.F. "The Effect of Physical and Mental Practice of Psychomotor Skills on Chemistry Student Laboratory Performance." *Journal of Research in Science Teaching* 16: 473–479; 1979.

30. Champagne, A.B., and Klopfer, L.E. *op. cit.*

31. Ausubel, D.P. *op. cit.*

Ausubel, D.P.; Novak, J.D.; and Hanesian, H. *op. cit.*

Novak, J.D.; Ring, D.G.; and Tamir, P. *op. cit.*

32. Gagne, R.M. *The Conditions of Learning.* Second edition. New York: Holt, Rinehart and Winston, 1970.

Gagne, R.M. *Essentials of Learning for Instruction.* Hinsdale, Ill.: The Dryden Press, 1974.

33. Chiappetta, E.L. "A Review of Piagetian Studies Relevant to Science Instruction at the Secondary and College Level." *Science Education* 60: 253–261; 1976.

Lawson, A.E., and Renner, J.W. "Piagetian Theory and Biology Teaching." *The American Biology Teacher* 37: 336–343; 1975.

34. Raghubir, K.P. "The Laboratory-Investigative Approach to Science Instruction." *Journal of Research in Science Teaching* 16: 13–17; 1979.

35. Hill, B.W. "Using College Chemistry To Influence Creativity." *Journal of Research in Science Teaching* 13: 71–77; 1976.

36. *Ibid.*

37. Cunningham, H.A. *op. cit.*

38. Wheatley, J.H. "Evaluating Cognitive Learnings in the College Science Laboratory." *Journal of Research in Science Teaching* 12: 101–109; 1975.

39. Hill, B.W. *op. cit.*

40. Wheatley, J.H. *op. cit.*

41. Pappelis, C.K.; Pohlmann, M.M.; and Pappelis, A.J. "Can Instruction Improve Science Process Skills of Premedical and Predental Students?" *Journal of Research in Science Teaching* 17: 25–29; 1980.

Pohlmann, M.M., and Pappelis, A.J. "Improving Process Skills Among College Non-science Majors with "Science, A Process Approach" Materials." *Journal of College Science Teaching* 6: 167–169; 1977.

42. Ophardt, C.E. "Development of Intellectual Skills in the Laboratory." *Journal of Chemical Education* 55: 485–488; 1978.

43. Pavelich, M.J., and Abraham, M.R. "Guided Inquiry Laboratories for General Chemistry Students." *Journal of College Science Teaching* 7: 23–26; 1977.

Pavelich, M.J., and Abraham, M.R. "An Inquiry Format Laboratory Program for General Chemistry." *Journal of Chemical Education* 56: 100–103; 1979.

44. Atkin, J.M., and Karplus, R. "Discovery or Invention?" *Science Teacher* 29: 45–51; 1963.

Lawson, A.E., and Renner, J.W. *op. cit.*

45. Pickering, M., and Crabtree, R.H. "How Students Cope with a Procedureless Lab Exercise." *Journal of Chemical Education* 56: 487–488; 1979.

46. Boud, D.J., and others. *op. cit.*

47. Ben-Zion, M., and Goldschmidt, Z. "Simulated NMR Spectrometry and Shift Reagents." *Journal of Chemical Education* 54: 669; 1977.

48. Hoskins, L.C. "Pure Rotational Raman Spectroscopy: A Dry-Lab Experiment." *Journal of Chemical Education* 54: 642–643; 1977.

49. Hefter, J., and Zuehike, R.W. "Computer Simulation of Acid-Base Behavior." *Journal of Chemical Education* 54: 63–64; 1977.

50. Cavin, C.S., and Lugowski, J.J. "Effects of Computer Simulated or Laboratory Experiments and Student Aptitude on Achievement and Time in a College General Chemistry Laboratory Course." *Journal of Research in Science Teaching* 15: 455–463; 1978.

51. Lahey, G.F.; Crawford, A.M.; and Hurlock, R.E. "Use of an Interactive General-Purpose Computer Terminal To Simulate Training Equipment Operation." Technical Report No. NPRDC-TR-76-19, ED. 119735. 1975.

52. Fitts, P.M., and Posner, M.I. *op. cit.*

53. Fraser, B.J. "Evaluating the Intrinsic Worth of Curricular Goals: A Discussion and an Example." *Journal of Curriculum Studies* 9: 125–132; 1977.

54. Clarke, J.A. "The Role of the Content and Structure of Curriculum Material in Cognition." *Science Education Research* 1973, pp. 119–141.

55. Herron, M.D. "The Nature of Scientific Enquiry." *School Review* 79: 171–212; 1971.

56. Shulman, L.S., and Tamir, P. "Research on Teaching in the Natural Sciences." *Second Handbook of Research on Teaching.* (Edited by R.M.W. Travers.) Chicago: Rand McNally, 1973. (Copyright American Educational Research Association.)

57. Herron, M.D. *op. cit.*

58. Hegarty, E.H. "Practical Curriculum Development: Scientific Enquiry in University Classes." *Research and Development in Higher Education* 2: 133–142; 1979.

59. Tamir, P., and Lunetta, V.N. "An Analysis of Laboratory Inquiries in the BSCS Yellow Version." *American Biology Teacher* 40: 353–357; 1978.

60. Chiappetta, E.L. *op. cit.*

Haley, S.B., and Good, R.G. "Concrete and Formal Operational Thought: Implications for Introductory College Biology." *The American Biology Teacher* 38: 407–412, 430; 1976.

Herron, J.D. "Piaget in the Classroom: Guidelines for Applications." *Journal of Chemical Education* 55: 165–170; 1978.

Lawson, A.E., and Renner, J.W. *op. cit.*

61. Atkin, J.M., and Karplus, R. *op. cit.*

Karplus, R. "The Science Curriculum Improvement Study—Report to the Piaget Conference." *Journal of Research in Science Teaching* 2: 236–240; 1964.

62. Strike, K.A. "The Logic of Learning by Discovery." *Review of Educational Research* 45: 461–483; 1975.

63. Ausubel, D.P. *op. cit.*

64. Lawson, A.E., and Renner, J.W. *op. cit.*

65. Atkin, J.M., and Karplus, R. *op. cit.*

Lawson, A.E., and Renner, J.W. *op. cit.*

66. Venkatachelam, C., and Rudolph, R.W. "Cookbook Versus Creative Chemistry." *Journal of Chemical Education* 51: 479–482; 1974.

67. Lerch, R.D. "An Evaluation of the Divergent Physics Laboratory." *Science Education* 57: 153–160; 1973.

68. Pavelich, M.J., and Abraham, M.R. *op. cit.*

69. McCaulley, M.H. "Personality Variables: Modal Profiles That Characterize the Various Fields of Science and What They Mean for Education." *Journal of College Science Teaching* 7: 114–120; 1977.

70. McCaulley, M.H. *op. cit.*

Rowe, M.B. "Who Chooses Science: A Profile." *Science Teacher* 45 (4): 25; 1978.

71. Charlton, R.E. "Cognitive Style Considerations for the Improvement of Biology Education." *American Biology Teacher* 42: 244–274; 1980.

72. Eggins, A. "The Interaction Between Structure in Learning Materials and the Personality of Learners." *Research in Science Education* (in press).

73. Gagné, R.M., and White, R.T. "Memory Structures and Learning Outcomes." *Review of Educational Research* 48: 187–222; 1978.

74. Holliday, W.G. "The Effects of Verbal and Adjunct Pictorial–Verbal Information in Science Instruction." *Journal of Research in Science Teaching* 12: 77–83; 1975.

75. Holliday, W.G., and Harvey, D.A. "Adjunct Labeled Drawings in Teaching Physics to Junior High School Students." *Journal of Research in Science Teaching* 13: 37–43; 1976.

76. Gagné, R.M., and White, R.T. *op. cit.*

77. White, R.T. "Relevance of Practical Work to Comprehension of Physics." *Physics Education* 14: 384–387; 1979.

78. Durand, D.P., and others. *MELE (Meaningful Enjoyable Learning Experiences) in Microbiology.* Second edition. Iowa State University: Burgess, 1973.

79. Ramette, R.W. "Exocharmic Reactions." *Journal of Chemical Education* 57: 68–69; 1980.

80. Novak, J.D. "Applying Psychology and Philosophy to the Improvement of Laboratory Teaching." *American Biology Teacher* 41: 466–470, 474; 1979.

81. Novak, J.D. *op. cit.*

Stewart, J.; Van Kirk, J.; and Rowell, R. "Concept Maps: A Tool for Use in Biology Teaching." *American Biology Teacher* 41: 171–175; 1979.

82. Norman, G.R., and others. "Clinical Experience and the Structure of Memory." Paper presented at the Research in Medical Education Conference, 1979.

83. Gunstone, R.F., and White, R.T. "A Matter of Gravity." *Research in Science Education* (in press).

84. *Ibid.*

85. Adams, R.S., and Biddle, B.J. *Realities of Teaching: Explorations with Videotape.* New York: Holt, Rinehart and Winston, 1970.

86. Dunkin, M.J., and Biddle, B.J. *The Study of Teaching.* New York: Holt, Rinehart and Winston, 1974.

87. Parakh, J.S. "A Study of Teacher–Pupil Interaction in High School Biology Classes: Part II—Description and Analysis." *Journal of Research in Science Teaching* 5: 183–192; 1968.

Egelston, J. "Inductive vs. Traditional Methods of Teaching High School Biology Laboratory Experiments." *Science Education* 57: 467–477; 1973.

88. Parakh, J.S. *op. cit.*

89. Parakh, J.S. *op. cit.*

Balzer, L. "Non-Verbal and Verbal Behaviors of Biology Teachers." *American Biology Teacher* 31: 226–229; 1969.

90. Parakh, J.S. *op. cit.*

91. Parakh, J.S. *op. cit.*

Balzer, L. *op. cit.*

92. Parakh, J.S. *op. cit.*

93. Hegarty, E.H. "Levels of Scientific Enquiry in University Science Laboratory Classes: Implications for Curriculum Deliberations." *Research in Science Education* 8: 45–47; 1978.

Shymansky, J.A.; Penick, J.E.; and Kyle, W.C. "How Do Science Laboratory Assistants Teach?" *Journal of College Science Teaching* 9: 24–27; 1979.

Tamir, P. "How Are the Laboratories Used?" *Journal of Research in Science Teaching* 14: 311–316; 1977.

94. Hegarty, E.H. *op. cit.*

Shymansky, J.A.; Penick, J.E.; and Kyle, W.C. *op. cit.*

95. Hegarty, E.H. *op. cit.*

96. Orgren, J. "Using an Interaction Analysis Instrument To Measure the Effect on Teaching Behavior of Adopting a New Science Curriculum." *Science Education* 58: 431–436; 1974.

97. Balzer, L. *op. cit.*

Evans, T.P., and Balzer, L. "An Inductive Approach to the Study of Biology Teacher Behaviors." *Journal of Research in Science Teaching* 7: 47–56; 1970.

Power, C.N., and Tisher, R.P. "Relationships Between Classroom Behavior and Instructional Outcomes in an Individualized Science Program." *Journal of Research in Science Teaching* 13: 489–497; 1976.

Tisher, R.P., and Power, C.N. *The Effects of Teaching Strategies in Mini-Teaching and Micro-Teaching Situations Where Australian Science Education Project Materials Are Used.* Brisbane: Australian Advisory Committee on Research and Development in Education, 1973.

98. Power, C.N., and Tisher, R.P. *op. cit.*

99. Power, C.N., and Sadler, R. "Non-Linear Relationships Between Measures of Classroom Environments and Outcomes." *Research in Science Education* 6: 77–88; 1976.

Tisher, R.P., and Power, C.N. *The Effects of Classroom Activities, Pupils' Perceptions and Educational Values Where Self-Paced Curriculum Materials Are Used*. Monash University: Australian Advisory Committee on Research and Development in Education, 1975.

100. Tamir, P. *op. cit.*

101. Stake, R.E., and Easley, J.A., codirectors. *Case Studies in Science Education*. University of Illinois, Urbana–Champaign: Center for Instructional Research and Curriculum Evaluation and Committee on Culture and Cognition, 1978. Volumes I and II. (Prepared for National Science Foundation Directorate for Science Education, Office of Program Integration.)

102. Power, C. "A Critical Review of Science Classroom Interaction Studies." *Studies in Science Education* 4: 1–30; 1977.

103. Hegarty, E.H. *op. cit.*

104. Hegarty, E.H. "Practical Curriculum Development."
Tamir, P. *op. cit.*

105. Tamir, P. *op. cit.*

106. Kyle, W.C.; Penick, J.E.; and Shymansky, J.A. "Assessing and Analyzing Behavior Strategies of Instructors in College Science Laboratories." *Journal of Research in Science Teaching* 17: 131–137; 1980.

107. Clarke, J.D., and McLean, K. "Teacher Training for Teaching Assistants." *American Biology Teacher* 41: 140–144, 187; 1979.
Manteuffel, M.S., and Von Blum, R. "A Model for Training Biology Teaching Assistants." *American Biology Teacher* 41: 476–479, 491; 1979.
Marsi, K.L. "Mid-Term Evaluation of Teaching Assistants." *Journal of Chemical Education* 55: 574–576; 1978.
Renfrew, M.M., and Moeller, T. "The Training of Teaching Assistants in Chemistry: A Survey." *Journal of Chemical Education* 55: 386–388; 1978.

108. Lawson, A.E., and Renner, J.W. *op. cit.*

109. Karplus, R., and others. *Science Teaching and the Development of Reasoning: A Workshop*. Berkeley: The Regents of the University of California, 1977.

Chapter 8

1. Driver, R., and Easley, J. "The Representation of Conceptual Frameworks in Young Adolescent Science Students." Doctoral dissertation. Urbana: University of Illinois, 1973. (Unpublished.)

2. Champagne, A., and others. "Interactions of Students' Knowledge with Their Comprehension and Design of Science Experiments." Pittsburgh: Learning Research and Development Center, University of Pittsburgh, 1980.

3. Minstrell, J. "Conceptual Understanding of Physics Students and Identification of Influencing Factors." Seattle: Mercer Island School District, 1980. (Unpublished.)

4. Viennot, L. "Spontaneous Reasoning in Elementary Dynamics." *European Journal of Science Education* 1 (2): 205–221; 1979.

5. Clement, J. "Students' Preconceptions in Introductory Mechanics." Amherst: Cognitive Development Project, University of Massachusetts, 1980.

6. Champagne, A.; Klopfer, L.; and Anderson, J. "Factors Influencing the Learning of Classical Mechanics." Pittsburgh: Learning Research and Development Center, University of Pittsburgh, 1979.

7. McCloskey, M.; Caramazza, A.; and Green, B. "Curvilinear Motion in the Absence of External Forces: Naive Beliefs About the Motion of Objects." Baltimore: Johns Hopkins University, 1980. [*Science* (in press).]

8. Arons, A. "Thinking, Reasoning, and Understanding in Introductory Physics Courses." *Proceedings of the GIREP Conference*. Rehovot, Israel: Weizman Institute, 1979.

9. Trowbridge, D. "An Investigation of Understanding of Kinematical Concepts Among Introductory Physics Students." Doctoral dissertation. Seattle: University of Washington, 1979.
Trowbridge, D., and McDermott, L.C. "An Investigation of Student Understanding of the Concept of Acceleration in One Dimension." *American Journal of Physics* (in press).
Trowbridge, D., and McDermott, L.C. "An Investigation of Student Understanding of the Concept of Velocity in One Dimension." *American Journal of Physics* 48 (12): 1020–1028; 1980.

10. Trowbridge, D., and McDermott, L.C. "An Investigation of Student Understanding of the Concept of Acceleration in One Dimension."

13. McDermott, L.C.; Piternick, L.; and Rosenquist, M. "Helping Minority Students Succeed in Science; I. Development of a Curriculum in Physics and Biology; II. Implementation of a Curriculum in Physics and Biology; III. Requirements for the Operation of an Academic Program in Physics and Biology." *Journal of College Science Teaching* 9: 135–140, 201–205, 261–265; 1980.

14. Δv = velocity, Δt = time, $\Delta v/\Delta t$ = acceleration.

15. McDermott, L.C.; Piternick, L.; and Rosenquist, M. *op. cit.*

16. Minstrell, J. *op. cit.*

17. Rosenquist, M., and McDermott, L.C. *Module III: Concepts of Motion.* Seattle: Physics Education Group, Department of Physics, University of Washington, 1980.

18. The investigation of conceptual understanding in kinematics was conducted by the Physics Education Group in the Physics Department at the University of Washington. This research was partially supported by the National Science Foundation under Grant Number SED78-17261.

Besides D. Trowbridge and the author, several other members of the group made important contributions that are gratefully acknowledged. D. Bartholomew, R. Lawson, and M. Rosenquist participated in the collection and analysis of data, and E.H. van Zee assisted in the preparation of this chapter. The cooperation of J. Rigden, Editor of the *American Journal of Physics*, is very much appreciated in granting permission to use excerpts from "An Investigation of Student Understanding of the Concept of Velocity in One Dimension" and "An Investigation of Student Understanding of the Concept of Acceleration in One Dimension," both by D. Trowbridge and L.C. McDermott.

Chapter 9

1. The number of researchers engaging in the task of understanding conceptual development is growing at a significant rate. This chapter was coordinated with Chapter 8 by McDermott, so the list of references following Chapter 8 is actually relevant to both chapters. I would especially like to acknowledge the work of A. Arons, J. Clement, J. Easley, L. McDermott, and D. Trowbridge. Their ideas have influenced me greatly. Finally, I wish to acknowledge the fact that collection of the research data reported in this chapter was made possible by Grant Number SED79-12824, with joint funding from the National Institute of Education and the National Science Foundation.

2. Trowbridge, D. "An Investigation of Understanding of Kinematical Concepts Among Introductory Physics Students." Doctoral dissertation. Seattle: University of Washington, 1979.

3. Bates, G.C. "The Role of the Laboratory in Secondary School Science Programs." *What Research Says to the Science Teacher.* (Edited by M.B. Rowe.) Washington, D.C.: National Science Teachers Association, 1978.

4. Inhelder, B., and Piaget, J. *The Growth of Logical Thinking from Childhood to Adolescence.* New York: Basic Books, 1958.

5. Minstrell, J. "An Evaluation of the University of Washington Program in Physical Science and Science Teaching for Elementary School Teachers." Doctoral dissertation. Seattle: University of Washington, 1978.

6. Rowe, M. B. "Wait-Time and Rewards as Instructional Variables: Their Influence on Language, Logic, and Fate Control." Paper presented at the Annual Meeting of the National Association for Research in Science Teaching, 1972.

Chapter 10

1. National Science Foundation, Science Education Directorate. *Projects in Science Education Development and Research.* SE79-80. Washington, D.C.: National Science Foundation, February 1979.

Atkinson, Richard C., and Lipson, Joseph I. "Instructional Technologies of the Future." Paper presented at the 88th Annual Convention of the American Psychological Association, Montreal, September 3, 1980.

2. Edwards, Judith B. "CAI and Training Needs in Professional Development and Educational Technology." 1980. pp. 117–124.

3. Up-to-date information and advice can be obtained from:

a. Dr. Judith B. Edwards, Northwest Regional Educational Laboratory, Portland, Oregon.

b. Dr. Ludwig Braun, College of Engineering and Applied Science, Department of Technology and Science, State University of New York, Stony Brook.

c. Dr. Dorothy Deringer, Science Education Development and Research Division, National Science Foundation, Washington, D.C. 20550.

4. Nelson, Theodore. *Computer Lib.* Chicago: the author, 1974
The Practitioner—A Newsletter for the On-Line Administrator 6 (1), October 1979.
Hands On!—A Forum for Science and Technology Educators. Spring 1980.
"Getting Started in Microcomputers." *Instructional Innovator* 25 (6), September 1980.
Purser's Magazine, Spring 1980.
5. Papert, Seymour. *Mindstorms: Children, Computers, and Powerful Ideas.* New York: Basic Books, 1980.
6. Robert Purser, editor. "Three Mile Island." *Purser's Magazine,* Spring 1980. p. 27.

Additional Reading
Lopez, Antonia. "Microcomputers and Education: Suggested Reading Material." *Creative Computing Magazine,* March 1981.
Classroom Computing News (P.O. Box 366, Cambridge, Mass. 02138).
Evans, Christopher. *The Micromillenium.* New York: Viking Press, 1979.
Personal Computing (70 Ninth Street, Peterborough, N.H. 03458).

Carl F. Berger is Associate Dean of the School of Education, University of Michigan. He was worked with the Science Curriculum Improvement Study (SCIS) at the University of California, Berkeley. Dr. Berger has done extensive research on science attitudes and their effects on learning, and he is currently involved in microcomputer studies. Much of his earlier work is focused on elementary and middle school concerns.

Ted Bredderman is on the faculty of the State University of New York, Albany, and was formerly an elementary science teacher. He is the author of numerous papers on science curriculum subjects and has completed a secondary analysis of research on the results of experiential learning. This study was part of a grant from the Research in Science Education (RISE) program for the National Science Foundation.

David P. Butts is Chairman of the Science Education Department at the University of Georgia. His principal work has been with elementary teachers and in science curriculum development as well as research. Dr. Butts is a former editor of the *Journal of Research in Science Teaching* and has been the recipient of the Robert Carlton Award of the National Science Teachers Association.

Drew Christianson is on the faculty of The Jesuit School of Theology at Berkeley, California. His speciality is philosophy and ethics. While he was with the Woodstock Theological Center in Washington, D.C., he led seminars for college and high school students concerned with issues of interaction among technology, business, government, and society.

Elizabeth H. Hegarty is a lecturer in microbiology at the University of New South Wales at Sydney in Australia and a Staff Associate of the World Health Organization Regional Teacher Training Center (Western Pacific Region) in the same city. Her teaching responsibilities are for science and medical students and higher degree candidates in Health Personnel Education. She has acted as consultant at courses on practical microbiology for high school teachers and has organized educational workshops for numerous groups of university teachers and health personnel.

David W. Johnson is a Professor at the University of Minnesota, Minneapolis. He is the author of twelve books and has published extensively in leading journals and professional studies. Dr. Johnson has worked with minority students in developing their reading skills in Indiana, New York, and Minnesota, as well as being a teacher and evaluator of a Freedom School conducted by the Harlem Parents Committee to teach Black History. Over the past several years he has received 11 research grants from such organizations as the National Science Foundation, Bureau of Education of Handicapped, and National Institute of Education.

Roger T. Johnson is an Associate Professor at the University of Minnesota, Minneapolis. A former elementary school teacher, he has also taught at the University of California, Berkeley. Dr. Johnson is the author of numerous books, curriculum programs, reports, journal articles, and research papers, and has received several research grants. He has served on various committees and task forces for environmental education and other science education concerns.

John J. Koran, Jr., is Professor and Curator of Education at the Florida State Museum. A former high school biology teacher in California, Dr. Koran has also taught at the University of Texas.

Joseph I. Lipson is with WICAT in Utah working on the uses of video-discs in education. He has been Division Director of Research and Development in the Educational Directorate of the National Science Foundation, and was formerly with the University of Pittsburgh.

Laurette F. Lipson is Chairperson, Department of Educational Services, at Baylor University in Texas. At Baylor she is directing a CAI system with special attention to minimizing the problems that exist for many people as a result of fear of computer terminology. She also has directed a large PLATO project.

Lillian C. McDermott is Professor of Physics at the University of Washington. She has received a Distinguished Service Citation from the American Association of Physics Teachers and formerly taught at City University of New York as well as Seattle University. Dr. McDermott has worked extensively in elementary teacher training, and she has been deeply involved in working with minority students at the college level. She has served on the Editorial Boards of both *The Physics Teacher* and the *Journal of College Science Teaching*.

James Minstrell is a high school science teacher on Mercer Island in the State of Washington. Dr. Minstrell received a grant from the RISE program of the National Science Foundation to support his work on the teaching and learning of physical science concepts, and has been doing the research in his own classroom.

James R. Okey is on the faculty of the University of Georgia and has a long history of involvement in curriculum development. Dr. Okey is a former president of the National Association for Research in Science Teaching, and much of his research has been focused on teaching at the secondary level.

Lynn Dirking Shafer is on the Educational Programming Staff at the Florida State Museum. She has had extensive experience teaching young people. For some time she has presented papers and has done research on learning in natural history museums as well as in other formal and informal settings.